LEARNING
TO LOVE

LEARNING
TO LOVE

HARRY F. HARLOW
UNIVERSITY OF WISCONSIN

ALBION PUBLISHING COMPANY
SAN FRANCISCO

ALBION PUBLISHING COMPANY

1736 STOCKTON STREET
SAN FRANCISCO, CALIFORNIA 94133

Library of Congress Catalog Card Number 75-182827
ISBN 0-87843-606-5

TO PEGGY

PREFACE

This short book is an attempt to present some of our ideas on the five forms of love. In addition, we contrast the positive emotions with a chapter on fear and anger and then conclude with some general considerations of social behavior.

The first of the affectional systems is maternal love and this is followed by infant love. These two forms of love are highly reciprocal, developing out of intimate maternal body contact in early infancy. The third love system described is age-mate or peer love of child for child, adolescent for adolescent. In many ways, peer love is the major determinant of all subsequent social and sexual development. Peer love is crucially important for the development of the fourth affectional system, heterosexual love. Peer passions, learned before sexual maturity guide the proper development of the heterosexual system. Paternal love is described as the love of the adult male for his offspring. This system provides a variety of social functions such as regulating infant play when it becomes too violent or inhibiting undue aggression in angry mothers.

Our research on this crucial area of affectional feeling for others continues and will be reported further in due course.

HARRY F. HARLOW

CONTENTS

CHAPTER ONE

THE AFFECTIONAL SYSTEMS

Whatever love may be, even the aloof scientist finds it difficult to approach the study of the phenomenon with total objectivity. Nevertheless, if we are to enter into the ordeal of affectional objectivism, we must recognize that love is not a single, invariant state, but that there are at least five basic kinds of interactive, interpersonal love. Let us define *love* as affectional feelings for others, thus ruling out self-love, or narcissism. Narcissism will be discussed later in terms of its function as an ego-protective mechanism.

The first of the affectional systems is maternal love, the love of the mother for her child. The second is infant love, the love of the infant for the mother, which can also be termed infant-mother love. The third is peer, or age-mate, love, the love of child for child, preadolescent for preadolescent, and adolescent for adolescent. This love system may, with good fortune, last throughout a first marriage, and with unusually good fortune it may even survive a second marriage. The fourth love system, heterosexual love, is one in which age-mate passion is augmented by gonadal gain. In deference to the two sexes, it is the period in which androgenic anxieties are quieted by estrogenic ecstasies. Even though adequate heterosexual acts may be, and have been, achieved unamelio-

LOVE

rated by love, the fact remains that adequate antecedent age-mate love is a prerequisite for heterosexuality, certainly in primates, probably in dogs [Beach, 1969], and possibly in male guinea pigs [Louttet, 1927]. Thus we must clearly concede the existence and remarkable motivational power of romantic love, the hallucinatory network that surrounds the image or form of the loved one. Heterosexuality as a complete love system appears to require memory as well as mating.

The fifth love system is that of paternal or father love, the love of the adult male for his family or members of his social group. Paternal love is expressed primarily in terms of protective functions, but it may also appear in more subtle forms, such as play with children. Paternal love may be lavished on a specific female and her offspring, as in the case of the titi monkey, the gibbon, and some nineteenth-century human males, or it may be very diffuse, as in rhesus and howler monkeys, chimpanzees, and the heroes of the Hollywood heyday.

Our description of five separate and discrete love systems is not meant to imply that each system is physically and temporally separate. Actually, there is always an overlap, so that affectional motives are continuous as the different forms and facets of love evolve. As with other aspects of development, each love system prepares the individual for the one that follows, and the failure of any system to develop normally

3

deprives him of the proper foundations for subsequent increasingly complex affectional adjustments. Thus the maternal and infant affectional systems prepare the child for the perplexing problems of peer adjustment by providing him with basic feelings of security and trust. Playmates determine social and sexual destiny, but without the certain knowledge of a safe haven, a potential playmate can at first sight be a frightening thing. By the same token, age-mate experience is fundamental to the development of normal and natural heterosexual love, whether this passion is deep and enduring as in most men or trivial and transient as in most monkeys. In all primates the heterosexual affectional system is hopelessly inept and inadequate unless it has been preceded by effective peer partnerships and age-mate activities.

The ties formed during each of the affectional systems are so strong and binding that they may sometimes impede transition to the appropriate new system when it eventually matures. Freud and others have described this difficulty as *fixation* at some infantile or early level. Thus an inability to break the maternal bond may hinder the development of peer affection, or some same-sex peer tie may block later heterosexual acceptance. Fortunately the very process of affectional maturation serves to offset affectional fixations. When maturation fails we may have to resort to learning—and when learning fails we may have to resort to psychiatry.

Although love has been an almost exclusive preoccupation of literature and legend, objective reports on human love are conspicuously scarce. Perhaps psychologists can live without love, and if so, this is doubtless the destiny they deserve. Fortunately for the rest of us, our meager data on man are supplemented by a wealth of data from other sources. The most relevant material, not just for the study of sex, but also for the study of love in all its manifestations, has been cooperatively provided by the monkey. Of course monkey data do not give a total picture of human devotion. Monkeys are much simpler than people. However, for this very reason they give us a clearer picture of the basic love systems, the nature of the variables underlying each, and the problems and perils of transition from one system to another. Furthermore, the scanty human experimental data, as well as the relatively rich human clinical data, can be seen in best perspective within the factual framework of primate affection.

MATERNAL LOVE

Mother love is one of the two love systems which has not remained unwept, unhonored, or unsung. Mother love has attracted the radiant raptures of poets and painters. Poets since Solomon have composed sensuous sonnets on the blessed beauties of the maternal face, form, and functions. Artists since time immemorial have immortalized the madonna, generally in a manner that in no way left her anatomical capabilities to the imagination. Scientists have also written about mother love. Some of their theses are true, and some are not.

4

The first scientific essay on mother love was strictly Freud [1949];
many subsequent essays have been strictly fraud. In tracing psychiatric
case histories as far back as possible Freud eventually discovered motherhood, a discovery as important as that made by Columbus, and equally
inevitable. Since neither he nor any other man really succeeded in going
back farther than birth, Freud concluded that motherhood was his
long-sought love and the ultimate and only source of all affection. One
analyst did believe he could go back to the womb, but this was Rank
[1952] folly. Freud conceived of the mother-child relationship as en-
compassing all the forces that shape the adult personality, with malad-
justments at this level the sole cause of any later emotional catastrophes.

Although we would never discount the importance of mothers,
it now appears that mother-child attachments are actually less critical
to adult social and sexual relationships than the interactions arising from
age-mate or peer attachments. Be that as it may, it is meaningless, and
even misleading, to judge the importance of the various affectional systems in terms of the end result. Each system evolves from the one that
precedes it, and the faulty development of any system, or the faulty
transition from one system to another, may arise from any number of
variables.

Since the affectional systems develop in sequence, let us examine
the mechanisms and their interactions in the same sequence. The chief
factor in maternal capabilities is a variable which is innate and genetically
determined—being born a girl. Females have an enormous advantage in
becoming effective mothers, since they will eventually possess the physical and physiological characteristics needed for effective motherhood—
the happy triumvirate of hope, health, and hormones. However, there
are more subtle factors than those built into the body and borne by the
blood that give females an enormous advantage in mastering the meaning of motherhood. There may actually be inherent differences in the
central nervous system.

Recent research efforts have been directed toward examining sex
differences in human newborns, differences that occur so early they
cannot be attributed to learning. On one hand, newborn females seem
to be more responsive to skin exposure. They react more strenuously
than newborn males to a covering blanket, are more disturbed when
their skin is stimulated by an air jet, and are higher in basal skin con-
ductance. On the other hand, newborn males raise their heads higher
than newborn females [Hamburg and Lunde, 1966]. Behavioral differences between the sexes have often been attributed entirely to learning
and culture. Learning is no doubt a factor of considerable significance,
particularly in human beings, but there are also far more subtle, secre-
tive, inborn variables. By the time children are four or five years old
they show awareness of sex-appropriate behavior which appears to be
determined by a combination of genetic factors and experience [Gari
and Scheinfeld, 1968].

As we shall see in Chapter 2, sex differences in nonhuman primates cannot be explained simply in terms of learned variables. There is

5

FIGURE 1–1   *Elicitor of female ecstasy response.*

reason to believe that genetic variables condition similar differences in human primates. The gentle and relatively passive behavior characteristic of most little girls is a useful maternal attribute, and the more aggressive behavior of most little boys is useful preparation for the paternal function of protection.

In addition to differences in general personality traits, there are sex differences more directly related to motherhood. Girls will respond to babies—all babies—long before they approach adolescence. Attitude differences that are apparently inherent were demonstrated in the responses of preadolescent female and male macaque monkeys to rhesus monkey babies [Chamove et al., 1967]. Since the preadolescent macaques had never seen infants younger than themselves and had not been raised by real monkey mothers who could have imparted their own attitudes toward babies, we may assume that the differences in response pattern were primarily innately determined. When they were confronted with a baby monkey, almost all the responses made by the female monkeys were positive and pleasant, including contact, caressing, and cuddling. These maternal-type baby responses were conspicuously absent in the males. Instead the male monkeys exhibited threatening and aggressive behavior toward the babies, although fortunately this behavior never progressed to the point of real physical abuse or injury.

Some years ago the photograph shown in Fig. 1-1 was projected on a screen at a women's college in Virginia. All 500 girls in the audience gave simultaneous gasps of ecstasy. The same test has since been conducted with many college audiences. Not only are all-male audiences completely unresponsive, but the presence of males in coeducational audiences inhibits the feminine ecstasy response. Evidently nature has not only constructed women to produce babies, but has also prepared them from the outset to be mothers.

Maternal love differs from the other love systems in that the full sequence is normally recurrent. The love of the infant for the mother is normally nonrecurring. Peer companionships may come and go, but the total developmental cycle is never reinstituted. Heterosexual love may be fixated or fleeting—and with each new generation it seems to have become progressively more fleeting and less fixated—but the long-term developmental stages of heterosexual love are nonrecurring. With mater-

6

nal love, however, the advent of each infant initiates the entire sequence anew. In its early stages maternal behavior is based almost entirely on the mother's responses to the infant. As the infant matures and develops his own responses, these maternal responses are modified by the interaction between mother and child. Thus the mother's expression of love for her infant changes and in fact appears to follow some type of development pattern. Human mothers feed, fondle, and dress their infants for the first year. After patiently retrieving dropped toys, and even deliberately thrown toys, and placing spoons in fumbling fists, they gradually wean their infants from total dependence and start them on the road to independence.

In the monkey mother the appearance of each and every infant initiates the first of three stages of mother love—a stage of care and comfort, a stage of maternal ambivalence, and a stage of relative separation [Harlow et al., 1963]. These stages are not discrete or all-or-none in character; rather, each merges into the next. It is impossible to define exact temporal periods for any maternal affectional stage in either monkey or man, since there are large individual differences and variables of experience play an increasing role in the later stages. In general, however, the stage of care and comfort in monkeys evolves into the stage of maternal ambivalence when the monkey infant is about five months old, and the stage of maternal ambivalence evolves into the stage of maternal separation four months to a year later.

THE STAGE OF CARE AND COMFORT

The response of a human mother to her newborn depends on many personality and cultural variables and reveals a complexity of concerns. If the infant is one that she wanted, she may view it with love and affection and even consider it beautiful—truly a triumph of mind over matter. However, if she has any doubts or conflicts about her anguished achievement, her response may be one of complete indifference. Many human mothers typically have little maternal feeling until their infants have matured to the point that they can interact with their mothers by means of vocal and facial responses. When the infant begins to coo and smile in response to maternal vocalizations or manipulations, the mother responds with maternal love.

During the stage of care and comfort a primary function or obligation of the monkey mother is to provide her infant with intimate bodily contact, which is the basic mechanism in eliciting love from the neonate and infant. For almost all monkey mothers the initial appearance of the infant releases love, as evidenced by positive and spontaneous approach and cradling responses. During the first month of life the mother maintains her baby either in a close ventral-ventral position (Fig. 1-2) or in a looser and more relaxed cradling posture in which the infant's body is held gently within the confines of the mother's arms and legs. This position provides the maximum amount of bodily contact

7

FIGURE 1–3 *Maternal respon-siveness to twins.*

FIGURE 1–2   *Maternal ventral-ventral cradling of infant.*

between mother and infant. Such physical attachments seem to be comforting for both mother and infant, an observation that has given rise to the concept of *contact comfort*. The monkey mother both gives and receives contact comfort, and it may be assumed that contact is an important mechanism in eliciting maternal love from the mother.

The role of infant body contact, clinging, and nursing in eliciting maternal love was demonstrated some years ago at the Wisconsin Primate Laboratories. One monkey mother was separated from her infant and allowed to adopt a young kitten. Initially the monkey mother made every effort to mother the kitten, even to the point of initiating nursing. However, the kitten exhibited one enormous behavioral flaw; it could not cling to the mother's body. When the kitten failed to reciprocate the mother's contact-clinging efforts, monkey maternal love waned, and after 12 days the kitten was totally abandoned.

Another monkey mother that had been separated from her own infant was introduced to a month-old infant that had developed an autistic pattern of behavior in which it clutched and clung to itself while rocking and mouthing its penis. Whenever the potential foster mother approached and contacted this infant, it screeched violently and intensified its autistic actions. After several days the mother failed to show normal maternal responsiveness. When this mother was later offered a second infant, which was congenitally blind but would cling normally and nurse, her adoption of it was immediate and complete.

In a recent experiment [Deets, 1969] four monkey mothers were given the opportunity to adopt "twins," two newborns, neither of which was the mother's own infant (Fig. 1-3). All four mothers at first attempted to initiate ventral contact and suckling with their infants. However, the twin infants, clinging in a haphazard way to themselves as well as to the mother, did not provide the kind of clinging and suckling feedback that is maximally reinforcing for the mother monkey. The mothers experienced great difficulty in establishing a satisfactory contact-clinging relationship, and three of the four mothers became

8

ambivalent toward their twin infants and began to alternately accept and reject each. While the mothers eventually did accept the twins totally and permanently, affectional ambivalence continued for as long as a month in one case. Similar ambivalence was demonstrated by a monkey who gave birth to a pair of natural twins in our laboratory. These and many other observations lead us to conclude that an infant that does not feed back will not be fed.

Actually, during the infant monkey's first week of life maternal love is indiscriminate, and during this period a monkey mother will respond with equal intensity to her own or to a strange infant of similar age. Earlier research had indicated that mothers are very upset by separation from their infants, but that their distress is somewhat allayed when they can see the infants. However, a more recent experiment [Jensen, 1965] demonstrates that they are relieved by the sight of *any* infant until their own offspring are a week to ten days old. After that point only the sight of their own infants will satisfy them. Evidently a monkey mother develops feelings specific to her own infant only after she has interacted with it for a period of time.

A similar developmental process appears to take place in human mothers, at least with their first infants. Until the infant is four to six weeks old the mother perceives it as an anonymous asocial object and reports only impersonal feelings of affection. In the infant's second month of life, when it begins to exhibit visual fixation and following, smiling responses, and eye contacts, her maternal feelings intensify. She now begins to view the infant as a person with unique characteristics and believes that it recognizes her. By the end of the third month the maternal attachment is sufficiently strong that the infant's absence is experienced as unpleasant, and the imagined loss of the infant becomes an intolerable prospect [Robson and Moss].

A second primary obligation of mothers during the stage of care and comfort is that of satisfying the infant's biological homeostatic needs, particularly those of hunger and thirst, and some monkey mothers are admirably endowed along these lines (Fig. 1-4). Infantile satisfaction of hunger and thirst is essentially guaranteed from the monkey mother by the pattern of body-to-body and breast-to-breast ventral-ventral

FIGURE 1-4   *Madonna and child.*

LOVE: MATERNAL AND INFANT

positioning. The neonatal monkey is held tightly by the monkey mother, with its mouth at or near the level of the mother's breasts. Infantile attachments to the breast in both human and monkey infants are also facilitated by rooting reflexes, which are head-turning responses to cheek stimulation, and subsequent nipple attachment and suckling. Each monkey infant has an almost exclusive preference for one breast over the other; however, half the infants prefer the left breast and half prefer the right, demonstrating complete statistical impartiality.

Nursing is often achieved less flawlessly and faultlessly by the human mother, probably because of the intervention in the human of learned cultural factors. Possibly a little learning is a dangerous thing, and many people have trouble learning more. It is estimated that only about 25 percent of the mothers in the upper socioeconomic strata in this country nurse their babies. Some apparently do not choose to nurse their babies because they regard nursing as a crude, belittling, bovine act. Some human mothers have great difficulty in nursing or cannot nurse their babies because of abnormal nipple or breast structures [Gunter, 1961]. Nevertheless, the many studies comparing the fates of breast-fed and bottle-fed babies suggest that breast feeding in human infants is a variable of relatively little importance. While it appears to lead to no negative outcomes, at the same time few clearly positive effects can be attributed to the experience [Caldwell, 1964]. Breast feeding does provide occasions for intimate interaction between mother and baby, but it is probably the intimate interaction, not the source of milk, that is the critical factor. Many human mothers may be relieved to know that it is possible for them to elicit and maintain infant love when they are bearing nothing but a bottle. As long as the bottle at each feeding session is supplemented by tender loving care, the bottle-fed baby will achieve the same pervading love for the mother.

A final care-and-comfort task of tribulation, a test of maternal love and patience, is that of the control of the infant's eliminative functions. Countless monkeys have been raised on inanimate cloth surrogate mothers. The infants render these mothers a soggy mess in only a matter of hours. An amazing contrast, however, infant monkeys do not soil their real mothers. The techniques by which monkey mothers achieve this laudable goal remain a mystery, possibly to them as well as to us.

Toilet training of human babies, in contrast, entails a minimum of mystery and a maximum of agony. There are many techniques for early toilet training of human infants, and all have one thing in common—they do not work. Most studies show that eliminative functions cannot be trained reliably and effectively until the child has reached a certain level of maturation. Masochistic mothers are, of course, welcome to enter the ordeal as early as they wish. Some years ago, eleven months was the average age for the beginning of bowel training in the United States. Typically it takes about seven months to complete training, although some babies develop control within a few weeks, while others require a year and a half or more. Over the last three decades mothers have gradually postponed initiation of bowel training [Caldwell, 1964].

The more severe the training procedure and the earlier it is initiated, the more the child is upset by the training, with severity being the more important variable. A very substantial proportion of children with bowel and bladder malfunctions are found to have experienced very early or very severe training regimens, although no broad personality consequences have been convincingly demonstrated [Caldwell, 1964]. Nonetheless there is reason to believe that compulsive and impetuous maternal toilet training and its consequent storm and strife may be the primary factor in what Freud termed the anal-erotic stage of development. This period is probably entirely a product of culture and conditioning, with no truly unlearned biological base. Indeed Erikson [1950] contends that the underlying cause is a general conflict between the child's developing desire to gain some control over his own activities, as opposed to being controlled by others. In Western societies, where bowel and sphincter control are assigned great importance, they become the focus for this general conflict between self-regulation and regulation by others. In societies where eliminative control is not so highly valued, of course, this general conflict may become focused on some other area of behavior. Nevertheless, written references to the anal stage are so frequent that we must at least accord it lip service in passing.

The third obvious essential function of the mother at the maternal stage of care and comfort is that of protection, and this is dispensed by good mothers at a level of vigor and violence appropriate or more than appropriate to the situation (Fig. 1-5). Eventually every human mother sees her young son embroiled in more or less mortal combat with the neighbor's boy. Automatically maternal instincts provide her the obvious knowledge that it was the neighbor's boy who started the fight, and assuming the external and internal bodily patterns of the rhesus monkey mother, the human mother rushes out to ensure her child's salvation. Aggression matures less rapidly in females than in males, but aggression in both sexes reaches a similar frequency and ferocity of expression when the females become mature. This would make good evolutionary sense, since mature maternal females must

FIGURE 1–5   *Maternal protectiveness.*

exhibit aggressive responses to external animate objects that threaten their young.

The maternal protective function appears early in the initial maternal stage and continues long after this stage has passed. The protective function is commonly elicited during the stage of transition and ambivalence, with little or no decrease in belligerence. Actually maternal protective responses exist long after the infant has reached a state of relative physical and psychological separation from the mother. In their natural habitat monkey mothers are often observed to come to the aid and defense of their fully adult sons.

The protection of the infant, and to a lesser degree the juvenile and adolescent, is not an exclusive maternal function. In many primate species protection is in fact a prime obligation of the adult males, and it is a basic component of the paternal affectional system. Actually, the male primate is better endowed by form and fang to serve this function than is the female. In primate groups studied in natural or seminatural settings the adult males are observed to defend infants and juveniles against social dangers arising from outside the troop. Moreover, in some groups of some species the adult males help keep the infant out of harm's way within the troop, as in breaking up vigorous bouts of rough-and-tumble play that threaten to injure some of the younger and weaker participants [DeVore, 1963].

THE STAGE OF TRANSITION

The love of the infant for the mother is so strong and so enduring that the very survival of the species is dependent on multiple separational factors or forces. Many mothers sever the strings with apparent ambivalence, and this feat is not commenced with ease. The early independence training which encourages the child to reach, retrieve, and replace his own toys quickly disappears when a cry of pain is emitted. The mother rushes to the child and swoops him up with hands which are otherwise only extended for finger contact at street crossings. This is indeed a period when the mother plays the ambivalent role of sending her infants forth to fend for themselves while also darting forth to defend her darlings.

As we noted earlier, Lorenz [1952] observed that immediately after birth goslings followed the heavy-hipped waddling of their mothers automatically and compulsively, and later would not approach and follow any other animal. Furthermore, when he removed the mother from newly hatched goslings, the goslings followed Lorenz in the same compulsive manner. This attachment was so strong that when the geese were grown they would not respond positively to members of their own species.

Imprinting in the goose has been endowed with many mystical or mythical characteristics. How any greylag gosling ever escaped from the maternal bonds to subsequently procreate the species was never explained by Lorenz. A possible explanation is that the greylag goose became extinct millions of years before Lorenz observed them, but this is contradicted by the facts.

IMPRINTING

Imprinted on a greylag goose
Is trapped within a hangman's noose.
How does an infant break away
To be a goose himself some day?

There are, of course, documented cases of human children who never break away from overly strong maternal ties. The typical "sissy" is one more closely bound to his mother than to his peers. Occasionally we read of the mother-trapped man who becomes engaged to the girl of his choice at age twenty and then waits until his mother passes away 50 years later before he marries. There are psychiatrists who believe that some subtle something may be lost in these intervening years. Most "mamma's boys" eventually escape. However, Portnoy is not the only one to have complained [Roth, 1969] of the difficulty of doing so, and even in such pretended patriarchies as ancient Sicilian society, it is mamma who is the source of final appeal in prayer as well as practice.

In view of the obvious strength of maternal ties, there must be even more superordinate mechanisms that unshackle the child from maternal bondage at an appropriate age or stage, so that he can become an independently functioning individual. In primates there are at least three processes that operate to gradually weaken mother-infant ties— changes in the mother's responses to her infant, changes in the infant's responses to his mother, and changes in the infant's responsiveness to the stimuli of the outside world. We shall treat these three processes as components of a single mother-infant separation mechanism. Certainly all of them operate toward a common goal—to free the child from the mother.

During the peak of the care-and-comfort stage almost all the normal monkey mothers' responses are positive, including such acts as intense maternal ventral contact, infant cradling, nipple contact, and restraining and retrieving the infant as it gains mobility and tentatively begins to explore the outside world. After this point maternal warmth wanes and separation supervenes and succeeds. Undoubtedly this stating of maternal affection is in part a function of acts by the infant designed to achieve maternal liberation. By sixty to ninety days of age the monkey infant has developed competent locomotor skills and has become an animal that can and needs to explore all aspects of the world about him. However, the decreasing frequency of maternal affectional responses is predominantly a function of fading maternal feelings and functions (see Fig. 1-6).

The development of the transitional or ambivalent stage is indicated by changes in the nature and number of maternal responses, particularly restraining and retrieving. For example, during the early months the mother is responsible for most of the physical separations of mother and infant, but the transitional or ambivalent stage is best

13

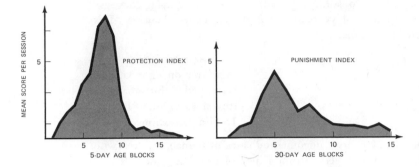

FIGURE 1–6   *Development of maternal protection and punishment.*

illustrated by the form and frequency with which the mother actively rejects her infant. Maternal punishment is almost nonexistent during the monkey infant's first three months of life. The frequency of negative maternal behavior—rejections, threats, and punishing—reaches a peak near the fifth month and then gradually declines to a low at about the ninth month. By that time either the monkey mother has faith in her infant or she has become terribly tired, while the infant has learned how to avoid maternal chastisement. Obviously the frequency of active maternal rejection is a joint function of both infant and maternal variables.

Rejection by the monkey mother begins as very gentle punishment, but increases in frequency and force as the infant matures. It eventually reaches the point at which the mother occasionally will deliberately dislodge the infant from her breast and body, or strike or stiff-arm the baby as it rushes impulsively to her in an attempt to gain maternal contact. As the transitional stage reaches a peak, maternal punishment is sometimes made without regard to any observable aggravation of the mother by the baby. However, the mother is ambivalent during this period. For example, if the infant appears to wander or stray beyond the maternal fold too quickly or too much, maternal responses appropriate to the stage of care and comfort reappear. Infantile independence during this transitional stage can be a source of frustration for the mother. She attempts simultaneously to protect and to emancipate. Human mothers no doubt experience similar ambivalence during this formative stage. Nevertheless, on all occasions in which the cherished infant is exposed to dangers, real or only imagined, the mother provides defense and protection with no hesitation or delay (Fig. 1-5).

Active rejection and punishment by the mother play an important role in helping the infant to break the overwhelmingly strong maternal ties and freeing him to make his own place in the world. Maternal rejection during this period is truly one of many forms of mother love; a mother who loves her infant will emancipate him.

Of course, in not all primate species do mothers rely on physical punishment to emancipate their infants. For example, chimpanzee mothers seldom punish their offspring. In the process of weaning their infants from the breast they use various techniques, such as tickling and

LOVE: MATERNAL AND INFANT

play, to distract the baby that attempts to gain access to the nipple [Van Lawick-Goodall, 1967]. Human mothers display an even greater variety of constructive techniques in seeking to aid the development of independence in their offspring. Frequently a toddler will be enrolled in a nursery school not solely for the mother's benefit, but to wean him from parental comforts and to expose him to new socially significant stimuli. Another approach is to provide toys that require the cooperation of peers, such as a teeter-totter, which can be teetered only by two.

THE STAGE OF RELATIVE SEPARATION

It is obvious that the affectional ties between mother and child are eventually dissolved—at least physically. In the case of both humans and monkeys the advent of a new baby is a variable of great importance, perhaps even primary importance, in establishing the stage of maternal separation. There are many mothers who have time for only one man at a time. However, the new baby only augments this natural process, and the older infant's own exploration and curiosity responses continue to function as a highly important force leading to separation.

The degree to which mothers and infants become physically and psychologically separated as the offspring mature varies greatly both within and among the different primate species. The rhesus monkey infants subjected to permanent physical separation from the mother eventually lose all affectional ties with the mother, and within one to two years ties with new associates replace the old bonds. In the original study on mother-infant separation [Hansen, 1966] the living quarters of the four mother-infant pairs tested were relatively small, and in these circumscribed surroundings mother and infant remained physically and psychologically close through the second year of life.

In large family-style living cages built to house nuclear family units (see Fig. 1-7) the first sibling may struggle successfully for many months

FIGURE 1–7   *The nuclear family unit.*

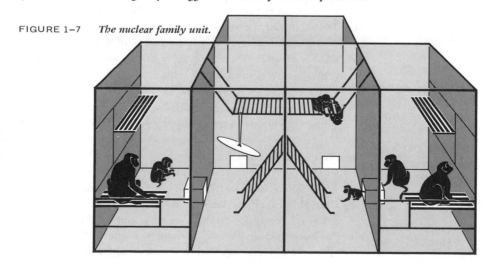

after the advent of a new sibling to maintain intermittent maternal contact, particularly at night. Maternal preoccupation with a new infant commonly augmented the detachment between the mother and the older infant. This was of little concern to the older infant during the day, when he spent his time in play with his age mates. However, as his interest in play and age mates departed with the daylight he commonly turned to prolonged and intensive efforts to regain maternal contact. On those occasions when even his most subtle and skillful efforts failed the rebuffed infant often tried to achieve close bodily attachment with some other mother and her offspring, and if even these desperate devices failed, he might as a last resort seek solace through ventral contact with his father, or even some other monkey's father. Although fathers are not totally adequate mother substitutes, apparently any body is better than no body at all.

Studies of two different species of macaque monkeys, bonnet and pigtail, have revealed striking differences in the extent of orientation toward the mother as the offspring matures. Adolescent and mature animals of both species spend approximately 80 percent of their time in friendly, positive interaction with other animals. However, in pigtail macaques this social interaction is directed almost exclusively toward the mother and close family members, whereas bonnets spend no more time interacting with kin than with nonkin [Kaufman and Rosenblum, 1969]. Cultural differences with respect to the extent of mother-infant separation are also apparent. For example, in the wild, in some groups of langur monkeys, after the second year of life the offspring interact with their mothers no more than with other adult females. In other groups of the very same species the offspring continue to direct their social interactions almost exclusively toward their mothers and close relatives well into adulthood. Similar cultural differences are evident within several other species of monkeys.

The human mother fortunately has both love to share and love to spare, for if she quickly separated physically and emotionally from her first infant upon the advent of the next, the problem of overpopulation would automatically be solved by the mortality rate among helpless first-born infants. Instead the human mother nestles and nurtures both, or even all, her offspring for a necessarily protracted time, thereby guaranteeing their survival—and along with it the full expression of sibling rivalry. Maternal separation in man is far more socially complex than it is in monkeys. It may be intermittent, semipermanent, or permanent throughout childhood as a result of baby-sitters, scholastic requirements, the miracle of remarriage, and in some instances deliberate choice. Maternal separation may be either physical or psychological, and it may be achieved by either maternal choice or offspring choice. Moreover, in man maternal contact is often maintained by written or spoken words long after physical separation has become a fact.

Physical separation between mother and child represents the final phase of each maternal-infant interaction sequence, and it is seldom, if ever, a simple process. Separation is commonly a period of stress, fear, and frustration, and the older infant must often be led to the top of the hill. Fortunately he almost always finds peers and playmates on the other side.

Infant love, which is the love of the neonate, baby, and young child for the mother, has often been confused with maternal love. These two affectional systems, the mother's love for the infant and the infant's love for the mother, operate simultaneously and are

INFANT LOVE

difficult to separate and unravel under normal physical and developmental circumstances. The intimate physical proximity and the close reciprocal relationship between mothers and infants make it difficult to distinguish the variables producing maternal affection for the baby from those producing baby affection for the mother. For example, nursing is an important variable in both affectional systems, but it is impossible under normal circumstances to determine its relative contribution to each. However, this does not mean that the mechanisms underlying each kind of love cannot be analyzed separately.

The mother has been physically aware of her infant's existence for some months while it was bundled in the pleasant paradise of the womb. Hence by the time of birth some form of maternal love, particularly in the human mother, is already in existence and is quickly affixed to the particular precious form which has made its appearance. Infant love, in contrast, is entirely indiscriminate at birth, and the neonate eagerly and equally attaches to any maternal object, animate or inanimate, that is endowed with adequate physical properties.

Inanimate stimulus models have been used for analyzing the variables underlying specific behavior patterns in fish and birds. The major advantage of using such models in research is the facility with which one or more variables can systematically be varied while all others are held constant. For example, models of the stickleback fish allowed Thompson [1963] to determine that the basic stimulus evoking aggression in the male stickleback was the belly color on another male stickleback.

Hess [1959], following the imprinting thesis of Lorenz, imprinted ducklings and chicks to inanimate wooden models and demonstrated the role of both visual and auditory variables in this limited learning process. He showed that the imprinting process depended in part on the amount of work or effort extended by the infant duckling. This differentiates imprinting from love; in love it is not the infant's effort, but the mother's that cements the affectional bond.

Similarly, the use of inanimate surrogate monkey mothers has allowed us to investigate important questions arising from several theories about the development of love and to analyze the individual stimulus properties of real mothers. For example, psychologists and psychoanalysts have long assumed that the infant's love for the mother developed

17

FIGURE 1-8 *Infant's religious conversion to diaper.*

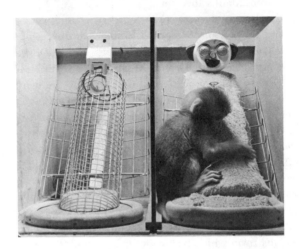

FIGURE 1-9 *Cloth and wire surrogate mothers.*

through association of the mother with organic pleasures resulting from the ingestion of milk and the alleviation of hunger. According to this "cupboard theory" of infant love, as Bowlby [1969] has termed it, the fundamental mechanisms are those related to functions of the breast.

Psychologists and their social science allies had long thought that mother love was a derived drive which developed from the association of the mother image with the alleviation of the primary drive of hunger. Motivation theory was dominated by the thesis that the only important unlearned motives were such homeostatic biological drives as hunger, thirst, elimination, and organic sex. All other motives were considered derived, learned, or secondary. Observations of monkeys suggest, however, that contact is actually the primary factor in the infant-mother relationship. Van Wagenen [1950], who raised newborn monkeys separated from their mothers, noted that unless their baskets were lined with soft cloth "even their feeding reflexes would become confused." The attachment of monkeys to their cheesecloth pads or diapers was so apparent that it looked as though they had joined a religious order (Fig. 1-8).

To determine whether infant love is learned or whether certain inherent properties of the mother elicit infant attachment Harlow [1958]

18

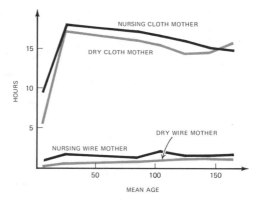

FIGURE 1–10   *Time spent on cloth and wire surrogate.*

constructed sets of nursing and nonnursing cloth and wire surrogate mothers. One primary variable differentiating the two types of surrogate mothers was that of body surface. The wire surrogates had bare bodies of welded wire and the cloth surrogates were covered by soft, resilient terrycloth sheaths. The two surrogates, shown in Fig. 1-9, had long, tapered bodies which could be easily clasped by the infant rhesus monkey. Some of the surrogates were endowed with a single breast, and some had none (the nursing surrogate mothers did not need two breasts, since none ever gave birth to twins).

In Harlow's classical dual mother-surrogate study a cloth mother and a wire mother were set up in a cubicle attached to the baby's living cage. For four newborn monkeys the cloth mothers lactated and the wire mothers did not; for four other monkey infants this condition was reversed. In both situations the infant received all its milk from the appropriate surrogate breast as soon as it was able to sustain itself in this manner. Total time spent in contact with the cloth and wire surrogate mothers under the two conditions of feeding is shown in Fig. 1-10. As the infants developed, those who had been provided with a lactating wire mother showed decreasing responsiveness to her and increasing responsiveness to the nonlactating cloth mother—a finding completely contrary to the interpretation that infant love is a derived drive in which the mother's face or form becomes a condition for hunger or thirst reduction. It is clearly the incentive of contact comfort that binds the infant affectionately to the mother. If derived drive is to be invoked as an explanation, this drive must be fashioned from whole cloth rather than whole milk.

There is seldom, if ever, any single cause for any behavioral act by any animal, particularly animals as complex as monkeys, apes, and men. Multiple stimulus variables operate to elicit each and every response, and there are commonly multiple antecedent variables, both maturational and learned. Hence, although these data show contact comfort to be the primary factor in the formation of infant-mother affectional bonds, it is by no means the only variable. To test the role of the breast as an affection-forming force infants were reared with two cloth surrogates, one with a functioning breast and the other with none. Under these circumstances all baby monkeys showed a significant preference

19

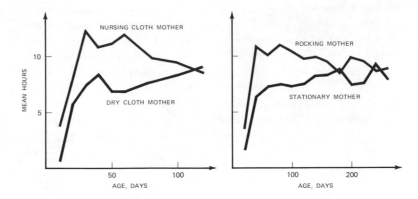

FIGURE 1-11   *Time spent on various cloth surrogates.*

for the mother with the bountiful breast (Fig. 1-11). In other words, activity associated with the breast and nursing is a significant variable when the more powerful variable of contact comfort is held constant. It has long been known that human babies are soothed by rocking motion, whether it is provided by the parent's nocturnal pacing with the baby in his or her arms or by a cradle. Studies also show (Fig. 1-11) that infant monkeys also have a preference for rocking surrogate mothers over stationary surrogate mothers [Harlow and Zimmermann, 1959].

In a series of small experiments designed to test the importance of temperature, since there is reason to believe that temperature is a variable of measurable merriment in the attachment of infant to mother, four infant macaques were given the choice of warm wire surrogates or cool cloth surrogates. The infants demonstrated a significant preference for the warm mother during the early days of life. This preference continued to be a significant variable until twenty days of age, when contact comfort became prepotent over temperature.

These follow-up studies illustrate the importance of history or experience as a factor in determining the role and importance of many affectional variables. It is difficult enough to determine the importance of the variables underlying infant love at any point, and since these variables change as a function of maturity and experience, the total task of variable analysis is heroic.

A similar multifactor theory concerned the human infant's love for and attachment to the mother. According to Bowlby [1969], the first attachment shown by the human infant to his mother is based on a number of primary, unlearned species-specific behavior patterns which he calls "instinctual response systems." These response systems mature and develop at different times and rates during the first year. Bowlby has described three such instinctual response systems in detail—suckling, clinging, and following—as mechanisms through which the infant actively maintains contact with his mother.

Two other instinctual response systems, ones in which the mother plays an active role, were mentioned in connection with the mother's

20

love for her infant. These response systems involve the infant's crying and smiling, which Bowlby suggests elicit maternal caretaking and strengthen the mother's attachment to her infant. At first the various instinctual response systems operate independently and lead the baby to attach himself to any mother figure; later, with development, these response systems become integrated and focused on a single mother figure. Although Bowlby's approach treats all infant love systems as merely component parts of one large, functioning love mechanism, his data add complexities rather than contradictions to the basic primate models we have discussed.

Recent research has shown that a number of stimuli and behaviors normally associated with the mother or maternal caretaking tie the infant to its mother. For example, Kessen and Mandler [1961] have demonstrated that the pure act of sucking, such as nonnutritive sucking on a pacifier, can in and of itself quiet an upset infant or prevent a content infant from becoming upset. Since these effects are observed in newborns before they are fed by breast or bottle, it is evidently determined by innate factors that are not learned as a result of hunger reduction following nursing. In monkeys bodily contact appears to be the primary variable determining attachment of the infant to the mother. However, certain forms of stimulation involving the distance receptors, particularly auditory and visual stimulation, apparently play a more important role in man than in monkey. For example, Salk [1960] demonstrated that the sound of the normal maternal heartbeat, as compared against other sounds, exerts a distress-relieving effect on the infant and even facilitates physical development. Other observations also point up the significance of distance-receptor stimulation in determining infant love in humans. However, studies of human infants exposed to intensive maternal handling and other forms of physical contact and infants whose maternal interaction was based primarily on auditory and visual stimulation, such as talking and smiling, indicate no significant difference in the infant-mother attachment [Schaffer and Emerson, 1964b].

Just as there is a progression in the stages of maternal love, so the infant love system proceeds through at least five stages—a stage of organic affection and reflexive love, a stage of comfort and attachment, a stage of security and solace, a stage of disattachment and environmental exploration, and a stage of relative independence.

THE STAGE OF ORGANIC AFFECTION

There has been prolonged and sometimes profound speculation by psychoanalysts, psychiatrists, anthropologists, and psychologists concerning the origin and nature of the baby's love for the mother and the stimulus or stimuli that call forth or evoke this attachment and emotional warmth. A common psychoanalytic position concerning the nature of the baby's original love is that the initial affectional force is not a love for the mother, since the neonate probably does not distinguish between his

21

own body and the body of any outsider, not even his mother's. Thus the original infant love is an egoistic love—not of himself or the body image of himself, but simply of organic sensations and satisfactions, initiated by the reflex act of nursing and maintained by the pleasures of food assimilation and the relief of organic tensions. This thesis seems reasonable, since approximately a half year elapses before the human infant differentiates between his own bodily self and the outside world of objects, animate or inanimate.

This period of lack of differentiation between internal sensations and precise external animate objects is probably greatly shortened in rhesus monkeys, because they are more mature at birth than human babies and develop about four times faster. Even so, a stage of organic affection probably exists. Since it is impossible to measure the introspective feelings of the neonate in either monkey or man, any early love or mere organic satisfaction must be based on inference rather than objectively measured fact. Almost by definition, however, a reflex stage of infant affection overlaps extensively, and probably completely, with an organic infant-mother affectional phase. Unlike the vague visceral feelings of a presumed organic stage, reflexes can be observed and objectively measured, and a reflex stage merits further consideration.

A basic law of growth and development is that both form and function begin at the head and sweep downward toward the posterior parts of the body. This cephalocaudal law of development holds for all species, but each species has its own characteristic pattern of growth. Opossums are born almost as soon as they are conceived and guinea pigs are nearly fully formed and functional at birth. In fact guinea pigs are approaching senility at birth; they are born strong of body and weak of mind.

The monkey is much more mature at birth than the human. Anatomically the monkey neonate is roughly equivalent to the human one-year-old, and its physiology and behavior are correspondingly advanced. Thus when the reflex stage of infant-mother love begins the human infant's reflexes are limited to the head and eyes, whereas the monkey infant has considerable reflex control over the trunk and legs as well.

In keeping with the cephalocaudal principle, human as well as monkey neonates possess some relatively efficient head and face reflexes. Sucking responses are adequately developed when and if the nipple is effectively attained. An allied reflex is the rooting reflex, which enables the neonate to search for the nipple by way of exploratory head movements. The normal adequate stimulus for the rooting reflex is stimulation of the neonate's cheek by the nipple. However, the rooting reflex may be elicited by stimulation of the monkey baby's cheeks with the human fingertip, preferably above and to one side of the oral cavity; the baby's head moves toward the direction of stimulation, and the lips clasp and try to engulf the stimulating object. The function of the rooting reflex is obvious; it aids both mother and neonate to achieve nipple placement and initiate nursing.

22

FIGURE 1–12   *Upward-climbing response.*

Another early response observed in human and monkey babies is vertical climbing, a response which appears to be meaningless, or even dangerous, but which must serve or have served some useful evolutionary purpose (see Fig. 1-12). Upward-climbing responses are observed in rhesus monkeys at a few days of age, since climbing responses in both men and monkeys mature in advance of walking and even more primitive locomotor patterns. If a neonatal monkey is placed head up at the base of an inclined plane which gradually rises to 2 or 3 feet above a tabletop or floor, the infant typically climbs slowly upward and, unless restrained or caught, crawls compulsively over the end of the plane and drops to whatever surface lies beneath—in all hope, the experimenter's hand. It is possible that this upward-climbing response serves to help position the infant monkey at or near the mother's breast. If he climbs forward and upward along the mother's belly wall, his movements will no doubt be restrained by the mother's arms at or near the level of the breasts.

Whatever the explanation, similar and equally malfunctional upward-climbing responses are frequently seen in human babies, who will try to climb stairs, chairs, sofas, and tables—and are saved only by adult intervention from steadfastly continuing right over the top and onto the floor. Even the painful experiences associated with falling from furniture usually fail to extinguish this innate response. As the human baby develops he sometimes finds new worlds to conquer, such as climbing onto and out of first-, second-, and even third-story windows, almost in direct refutation of his behavior in visual-cliff experiments. At this point the upward-climbing response becomes a response of limited survival value for the human.

The human and monkey babies differ very markedly in the development of the hands, arms, and trunk, which control the responses enabling the neonate or child to attach and maintain contact with the mother, primarily through groping, grasping, and clinging responses. The monkey possesses these reflexes at birth, and even before they become voluntary responses, he is able to obtain contact comfort from his mother largely by his own devices. Since it is five or six months before the human infant can grope for, grasp, and cling to his mother, for the first half

23

year he is completely at her mercy for contact comfort, the primary variable in the formation of deep and enduring infant-mother love.

Setting deadlines for primate maturational stages is mischievous and possibly meaningless. As we have already seen, infant organic love and reflex affection begin at the same time and may even wane and terminate at approximately the same time. The stage of comfort and attachment certainly overlaps them, but without question it continues after they have passed.

THE STAGE OF
COMFORT AND ATTACHMENT

The stage of comfort and attachment of the infant-mother affectional system begins as soon as the infant is able to attach to the maternal body and breast. The infant macaque commonly achieves this on the first day of life, and comfort and attachment is usually maintained into the second year. For the first six months of life or longer the human infant is totally dependent on the mother for contact and attachment, and human mothers vary greatly in the degree and amount of bodily contact given during nursing or through cuddling or simple play. Studies with monkeys indicate that with relatively limited affection, if it is given readily and consistently, an infant adjusts successfully within each period of affectional development and proceeds to the next. Minimal comfort and attachment are not recommended, but human mothers beset with other problems may find it comforting to know that limited affection is better than none.

During this stage contact comfort is clearly a primary variable in infant monkeys, and nursing factors, temperature, and proprioceptive variables are significant. It is unlikely that these variables differ in nature and importance in the human infant. However, as we have seen, exteroceptive variables are probably more important in man than in monkeys.

Another important behavior at this stage of infant development is primary object following. Bowlby suggests that primary object following is a powerful mechanism in attaching the infant to the mother, and possibly in producing security responses. Unfortunately there are very limited data on primary object following by either human or monkey children. From casual observation it appears that the rhesus infant follows maternal behavior and, within the limits of its capabilities, matches her actions. When the mother mouths and ingests a food substance, so does the infant; when the mother is startled or frightened, the infant clings to the mother's body for security and safety, while perhaps still continuing to observe. These behaviors enable infants to profit from maternal experience, so that their own exploratory behavior is not blind and the dangers inherent in untutored exploration are minimized. Furthermore, such maternal association, reinforced by maternal bodily contact during moments of doubt, and probably by other maternal variables as well, undoubtedly abet the development of the infant

24

not for the protective restriction of a real monkey mother or the laboratory environment. This fear of the strange or different develops gradually during the first three months of life.

A recent study has dramatically demonstrated the maturation of fear. Infants were reared from birth under conditions that isolated them from social stimuli, and pictures of monkeys engaged in various forms of behavior were then projected to them in their isolation chambers. One set of pictures depicted adult animals exhibiting facial threat grimaces. The infants in their isolation chambers showed little or no responsiveness to these threat-grimace pictures until they were about ninety days old, when the full fear pattern emerged suddenly and at peak strength [Sackett, 1966]. In a detailed study of this phenomenon in human children Schaffer and Emerson [1964b] found that under completely controlled experimental conditions the results were highly similar to those reported by Spitz. Pathological fear in the mother's absence appeared as early as six months and as late as fourteen months, and a few recalcitrant children showed no fear with the mother removed. Man is a strange animal and human data are intrinsically variable.

STAGE OF DISATTACHMENT AND ENVIRONMENTAL EXPLORATION

As the infant human and macaque mature, many factors arise that tend strongly toward maternal disattachment. One variable is maternal punishment, which, absent during the early months, peaks between the fourth and eighth month in the monkey, and drops to a low, stable baseline by about the tenth month. Monkey maternal punishment is usually gentle and restrained, but even when it is harsh and vigorous no mother has ever been seen to injure her own baby, or even attempt to do so.

Although maternal punishment is an important development at this stage, the infants are also lured by natural forces from maternal charms. There is general agreement that the primary mechanism in disattachment from the mother is maturation in the infant of freedom of activity, curiosity, and manipulatory needs concerning the external physical environment, and later, social drives for interaction with age mates and other members of the species. The primary maternal contribution is neither maternal restraint nor punishment, but rather the positive factor of establishing the personal and social security which the infant must have to be able to leave the mother and the mother's domain.

Disattachment in the human infant has been reported by Rheingold and Eckerman [1970], who studied human babies, some less than a year old and so young that they locomoted by wriggling or crawling. After the infants were placed in a room with their mothers, who doubtless provided them with effective maternal security, incentives rang-

solace and security stage. Such responses are essentially nonexistent during the first 60 days of monkey life and then increase in frequency during the next five months.

THE STAGE OF SOLACE AND SECURITY

The stage of solace and security in infant-mother affection develops at the time, or probably some time before, human or monkey infants develop affection for their specific mothers. Security at this stage is expressed by the willingness of the infant to wander beyond the mother's physical and functional confines to explore the strange new world of objects, playthings, and playmates. Early in the security stage the exploration is minimal and is always under watchful maternal eyes, with the infant frequently returning to the mother for care, comfort, and solace. As time passes the infant's expeditions are longer, with less frequent returns to the mother, and maternal reassurance may change from profound contact clinging to token maternal assurances expressed by a bare flick of the fingers or casual contact with the toes.

Some years ago, in an experiment planned to induce psychopathological behavior in infant monkeys, four surrogate monster mothers were constructed. One was a shaking mother which rocked so violently that the teeth and bones of the infant chattered in unison. The second was an air-blast mother which blew compressed air against the infant's face and body with such violence that the infant looked as if it would be denuded. The third had an embedded steel frame which, on schedule or demand, would fling forward and knock the infant monkey off the mother's body. The fourth monster mother, on schedule or demand, ejected brass spikes from her ventral surface, an abominable form of maternal tenderness and succor. All the monster mothers, however, had a comfort-giving cloth surface.

As disturbing as these monster mothers were, the infant monkeys did not even leave the bodies of the air-blast and rocking mothers, since the mother is an infant's only source of solace or succor, and the only response of an infant in distress is to cling more tightly to the mother. The infants had no choice about their departures from the throwing-frame mother and the brass-spiked mother. Nevertheless, crying and complaining, they waited for the frame to return to resting position and the spikes to retract into the mother's body and then returned to the ventral surrogate surface, expressing faith and love as if all were forgiven.

The infant monkeys apparently showed no apprehension between these trials of torture, and no cumulative terror effects. Although the experiment failed to achieve its original ghoulish goal, no experiment could have better demonstrated the power of any contact-comfort-giving mother to provide solace and security to her infant. These mechanisms clearly superseded any discomfort, disturbance, and distress, despite the fact that nociceptive stimulation is presumably prepotent.

Love, all kinds of love, is a mechanism or variable of almost unbelievable power.

Since all kinds of monkey mothers can instill solace and security in their babies' hearts and heads, it is no surprise that baby monkeys clasp and cling to soft cloth surrogates many hours a day. Moreover, they achieve security from the cloth surrogates at about the same age that they would achieve a sense of succor from their own real monkey mothers. However, a study by Hansen [1966] reveals that the security imparted by inanimate cloth mothers is less than that imparted by real mothers, at least during the first two months of life. Of course real mothers would be more efficient protective sources than cloth mothers if the baby monkey were being threatened by living predators. However, the laboratory is designed to keep predators, other than human ones, out.

The original study of the relative affection of infant monkeys for cloth and wire mothers, lactating and nonlactating, measured not only responsiveness to these four kinds of mothers in terms of total contact time, but also choice responses to each maternal type when the infant monkeys were frightened by the sudden appearance of a malevolent mechanical monster (see Fig. 1-13). Their responses to this awesome sight were measured in terms of surrogate contact time. The initial haven was a cloth surrogate for about 70 percent of the one-month-old-monkeys and 80 percent of the two-month-old monkeys. Many of the responses to wire surrogates were momentary and were quickly reversed, as if the terror-stricken baby had dashed blindly to any object in the immediacy of need, and finding the wire surrogate comfortless, quickly reversed its choice and sought the security of any cloth surrogate, lactating or nonlactating.

In other studies monkey infants had to circle around a vertical Plexiglas barrier guarded by a horrible mechanical monster to reach a cloth surrogate mother. Although they were greatly disturbed by this situation, all eventually braved passage around the monster to reach the mother, where they would snuggle against her and quickly become emotionally composed. After a number of these sessions the infants would rush to the mother, and once they had achieved sufficient maternal comfort and security, they would explore the previously fearful experimental chamber. Indeed, within a few more days, several brave infants actually approached and manipulated the monster that had previously left them prostrate. Finally one of them dismembered a monster and tore it to shreds, a perfect illustration of the achievement of solace and security.

The nature and function of the stage of security and solace is clearly demonstrated by the infant's bravery when the mother is present and his terror when the mother is removed. Various infant monkeys raised on cloth surrogates were placed in a strange test room, bare except for a few playthings and their cloth surrogate mothers. As soon as an infant spied his mother he rushed over and clung tightly; only then would he explore the chamber and play with the toys (see Fig. 1-14). When the same infant monkey was placed in the same room with the

FIGURE 1-13 *Response to mechanical monster.*

FIGURE 1-14 *Baby exploring chamber and manipulating toy.*

FIGURE 1-15 *Infant frozen in terror in absence of surrogate.*

cloth surrogate absent, he froze in terror (Fig. 1-15) and failed to show any emotional recovery throughout the entire experimental session.

Even more striking results were obtained with infants raised on lactating wire mothers. When the wire-weaned monkey babies surveyed the chamber they made little or no effort to go to their wire mothers, but instead threw themselves prone on the chamber floor, crying and grimacing all the time, or huddled against a chamber wall, rocking back and forth with their hands over their heads or faces. Evidently even infant monkeys can find no comfort in wire women.

Essentially identical data have been amassed on human children tested in the presence and absence of their mothers. Spitz [1946a] first noted the desperate fear of human children in a strange room when their mothers were absent; he termed this *eight-month anxiety* to call attention to the maturational age variable involved. This mother-loss anxiety appears two or three months earlier in some babies and several months later in others [Spitz, 1950].

Since all mothers of any one species are more alike than different, the formation of highly specific personal bonds for a particular mother is of necessity learned. Infant responsiveness that is specific to a specific mother does not reach fruition in the macaque infant until about the fourth month. By this time the macaque baby is able to discriminate his own mother from other mothers, and this discrimination is aided and abetted by fear responses in the presence of strange inanimate and animate objects. Both phenomena are dependent on the maturation of underlying neuroanatomical mechanisms. The newborn macaque shows no fear of any object, animate or inanimate, regardless of how strange, and he would without question expose himself to fatal dangers were it

ing from no toys to three toys were placed in a large adjacent room whose entrance shut off visual contact with the mother. The distance that the infant traveled away from the mother was positively related to the infant's age, a fact not entirely surprising. No apparent differences between sexes were disclosed, but there was a positive relationship between number of toys in the open field and the time spent away from the maternal security figure. Moreover, the infants went from the starting room to the open field and back, probably for maternal support, even when no toy was present.

In this study the mothers apparently remained totally passive and behaved more like our inanimate cloth surrogate mothers than like normal mothers. Thus, despite its merits, the study does not reveal the development of maternal changes during the stage of mother disattachment; disattachment and distance traveled would certainly be maximized, since maternal restraining variables were eliminated. Actually, such a stage does not normally occur in human infants until they are much older than one year. The investigators' conclusions, however, are equally apt as an interpretation of the maternal disattachment phase in both monkey and human infants [Rheingold and Eckerman, 1970, p. 78]:

> The infant's separating himself from his mother is also of psychological importance, for it enormously increases his opportunities to interact with the environment and thus to learn its nature. . . . Similarly, what can be learned about the physical environment parallels [what] can be drawn for the social environment.

STAGE OF RELATIVE INDEPENDENCE

The important variables relating to the infant's separation from the mother are similar to those discussed in terms of the mother's separation from the infant. However, there are apparently important differences between the infant-mother disattachment phase and the infant-mother separation stage. There are, moreover, important activities in this stage which are initiated by the infant, and it is probable that these activities are more important than those of the mother. Disattachment is an almost totally infant-guided process, with the mother's role remaining passive unless disturbance or danger develops, whereas separation appears to be primarily mother-determined, and even the most resolute infant struggles against it in moments of apprehension and whenever daylight fades into dusk.

In the next chapter we shall consider the development of the age-mate or peer affectional system. Probably the most important function of the maternal and infant love systems is to prepare the infant, human or monkey, to indulge in the wonders of age-mate acceptance and interaction.

29

SUMMARY

Love, or affectional feeling for others, may be described in terms of five basic systems, each of which provides a foundation for the increasing requirements of the next. The first affectional system is maternal love, the love of the mother for her child. The second is infant love, or infant-mother love. This is followed by peer, or age-mate love, which is fundamental to the development of normal heterosexual love. The fifth system is paternal love, the love of the adult male for his family or social group.

Mother love is the first affectional system experienced by the newborn. Maternal characteristics appear to some extent to be genetically determined, since behavioral differences between the sexes are apparent even in newborn infants. These differences increase with age, but not all of them can be attributed to learning. The maternal affectional system is the only one of the love systems in which the full developmental sequence is repeated; it is initiated anew with each successive infant. Mother love is indiscriminate, and in human mothers often absent, at the outset. It appears to be elicited by the infant's response to contact comfort provided by the mother.

The first phase of maternal love, the stage of care and comfort, is characterized by the infant's dependence on the mother for satisfaction of physical and emotional needs. The stage of transition is marked by intermittent and ambivalent rejection, as evidenced by a decrease in the nature and number of maternal responses to the infant. The final stage, of relative separation, varies greatly in different primate species and even among individuals. The advent of a new baby often augments separation of the mother and the older infant. However, the primary force leading to separation is normal maternal rejection during this stage and the growing infant's own motivation to explore his growing world.

Infant love is indiscriminate for a much longer period than mother love. In monkey infants this attachment appears to be related primarily to contact comfort rather than to result directly from association of the mother with satisfaction of the organic need for food. Of course other variables, such as the source of food, warmth, and rhythmic rocking motions, also play a role in this attachment. Certain types of stimulation, particularly stimuli involving distance receptors, may play a more important role in human infant attachment for the mother than they do in nonhuman primates.

The infant affectional system is characterized by five major stages. The stage of organic affection and reflexive love, presumed to be related to organic satisfaction from nursing, is indicated by the various reflexes in the infant monkey which ensure its clinging to the mother and finding the nipple. The stage of comfort and attachment begins as soon as the infant is able to attach itself to the mother; the most important variable in this stage is clearly contact comfort. The stage of solace and security develops around the time that infants form a specific attachment to a specific mother. This stage is marked by curiosity and exploration, but with a constant return to contact with the mother for solace and in many cases by gentle punishment, and in human mothers by distraction and other creative means. However, the primary mechanism in the infant's disattachment from the mother is the maturation of freedom of activity, curiosity, and manipulative needs in connection with the external physical environment.

The final stage in the infant love system is the stage of relative independence. In its most important phases the dissattachment of infant from mother is an almost totally infant-guided process, while separation is determined primarily by the behavior of the mother. By the time of separation the mother-love and infant-love systems have served their important function of preparing the infant for later age-mate interaction.

CHAPTER TWO

**AGE-MATE OR
PEER LOVE**

Probably the most pervading and important of all the affectional systems in terms of long-range personal-social adjustments is the age-mate or peer affectional system. This system develops through the transient social interactions among babies, crystallizes with the formation of social relationships among children, and then progressively expands during childhood, preadolescence, adolescence, and adulthood. Individual age-mate or peer affectional relationships may exist between members of the same sex or opposite sexes. There are, however, certain basic physical, biological, and behavioral differences conducive to sexual separation in infancy, and subsequently, to the development of intensified heterosexual interests and choices which begin in late adolescence.

The primary positive variable pervading peer love is that of play, which progresses from the asocial exploratory play characterizing early infancy to parallel play, and subsequently to the multifaceted forms of social, interactive play achieved by the child, the adolescent, and the adult. There have been many theories, and even some overinterpretations, of the social importance of play, but all have failed to see play as the

32

LOVE

primary factor in integrating the earlier forms of love, in fulfilling the inexhaustible social needs of the peer affectional or love system, and in organizing the experiences of the individual in preparation for the requirements of heterosexual or adult attachment.

Quaint and curious are some of the theories of early authors concerning the forms and functions of play. These speculations were based for the most part on anecdotal evidence of children's play, forgotten physiological theories, or an awakened interest in an evolutionary hypothesis long since put to rest. Spencer [1873] conceived of play as simply a release of surplus energy, as play must look to any aging man. The only relationship, of course, is that vigorous play and available animal energy are maximal at approximately the same age or developmental period. A theory of historical interest is that of Hall [1920], who, imbued with Haeckel's theory of evolutionary recapitulation, believed that each child reenacted the activities of his primitive ancestors and that ontogeny paralleled the phylogenetic evolution of man in all regards. Hall would be delighted to know that infants can swim before they can crawl, crawl before they can walk, and walk before they can speak [McGraw, 1935]. However, these fascinating facts have no importance in the development of human or monkey play. As we shall see, even the

33

development of play in infants of such closely allied species as macaque monkeys and men does not fit any such pattern as that of following the phyletic leader. There are basic qualitative differences in the play of men and monkeys at every stage and every age.

Groos [1901], on the basis of extensive personal observations of human and animal children, discovered and described the fact that the play of children had an essential functional value, and was not simply a diversion for the release of energy. Unfortunately he did not recognize the orderly progression of various developmental stages of play and failed to point out the effect of each stage on each subsequent stage. Instead, he saw infant play as direct training for future adult activity.

The child does learn and master such traits as affectional expression, friendship formation, social ordering, and even aggression in relations with age-mate friends and foes. However, these traits do not transfer to adulthood unaltered or unchanged. What is proper with peers is improper with parents. These basic behaviors may inherently influence the social development of all later stages, and the traits named are admittedly basic to the subsequent mastery of adult goals, games, and gambits. However, the infant plays for play itself, unaware and unconcerned about its adult significance. Indeed, as emphasized by Dewey [1922], play is freed from "the fixed responsibilities attaching to the special calling of the adult."

The data on human play are almost without exception limited, lonely, cross-sectional studies, observational rather than experimental in design, and conducted in homes and nursery schools, where control and management of salient variables is nearly impossible. All homes hope for happiness, and all nurseries need neatness and nicety. By the time the peer pattern has approached perfection or importance, nursery schools are only memories, and homes are temporary havens where vigorous play is strictly forbidden. In contrast, much can be learned in the laboratory, where an animal can be born, reared, and studied for years on end under scientifically controlled surroundings. In no other way can adequate information be obtained concerning the fundamental nature of play, the developmental interactions among the systematically sequencing play systems, the effects of enriching play development, and the heartrending catastrophes that transpire when normal play development is denied or distorted.

Although peer love and play have an important role in socialization, and possibly in some aspects of intellectual growth, neither peer love nor play arises from a vacuum or terminates in vacuity. Peer love is preceded by two antecedent love systems, mother love and infant love, which transmit their own heritage. It is also preceded by the two transitional mechanisms of contact acceptability and basic trust. Actually the transitional mechanisms may be of more social importance than the specific learnings acquired in the antecedent love system. It is possible that the love the infant has acquired for his mother, including the physical properties of her form and face, is transposed by simple stimulus generalization to the physical properties of peers. Such a theory is implicit in Freud's famous, and possibly fatuous, formulation of the power of the mother image in fashioning the wishes and wills, wondrous but not wanton, in each human adult male. Attempts by experimental psychologists to test such a theory have had doubtful results. Of course man may not be the best subject for such a test, but for some mysterious reason tests on other animals are rare. In casual observation the data on monkeys raised with surrogate mothers do not appear to support Freudian dogma. The first objects they found and loved in the world outside the maternal domain were inanimate rather than animate objects. Surely no real mother ever looked like a broken brickbat, a celluloid ball, or a multicolored painted piece of plywood.

Whether or not specific affectional attachments are passed on by way of stimulus generalization from mothers to peers, there are at least two very important transitional mechanisms for which we must account. The first of these transitional or developing mechanisms is the mutual acceptance of physical or bodily contact with members of the same species. We believe, in the absence of specific experimental data, that the intimate mother-infant contact comfort does transfer from the maternal figure to age mates or peers. Play in all its complex forms is impossible if bodily contact is looked upon as undesirable or loathsome. Just as there is seldom fun without feeling, there is seldom feeling without fun. A second overwhelmingly important positive transitional mechanism is the gift of basic trust and security, as described by Erikson [1950]. We have already discussed the formation of trust and security in connection with maternal and infant affectional systems. Here we simply accept reasonable basic security as an essential social antecedent to the formation of peer love. Without basic security, neither the human nor the monkey infant would be free to express physical contact and to explore the playthings and playmates that are essential for the formation of age-mate love.

The age-mate affectional system begins when the intimate physical bonds between mother and child weaken and the infant wanders beyond the range of the maternal body and the reach of the maternal arms. This detachment process is primarily a function of the infant's developing powers of curiosity and wanderlust, the only lust the buoyant baby really has. The act of detachment may be supported by maternal rejection and punishment or by maternal apathy and fatigue. Even human mothers become tired and troubled.

As we have seen, mother love is usually prolonged and persisting, although in some societies and some mothers it may be relatively brief. Moreover, its duration is frequently a function of normal or abnormal interaction on the part of mother or child. Infant love, which depends on specific mother recognition and attachment, is established at six to nine months. It wanes as dependency ties decrease, although its persistence depends on numerous social and situational variables.

The age-mate or peer system is less standard, in both its origin and duration, partly because of the increase in number and complexity of learned or cultural variables. It is theoretical folly to assign rigid age ranges for the affectional systems. Nevertheless, age boundaries have convenience value in planning and producing developmental research, normal or abnormal, and interpreting such data. With this injunction in mind let us say that the peer affectional system begins at about three years in humans, peaks between the ages of nine and eleven, and wanes with the onset of adolescence, when peer relations become hopelessly entangled with heterosexual affection. Similar developmental periods exist for the monkey, but they make their appearance much earlier because of the monkey's greater physical and physiological maturity at birth. Subsequent monkey developmental periods tend to be about one-quarter the duration of the analogous human stages. Thus monkey infant love is probably specific at one month, vigorous age-mate play is present at four months, and childhood ends at three years.

We have defined the affectional systems in terms of broad social developmental mechanisms, with suggested age ranges [Harlow and Harlow, 1965; 1966]. As a matter of convenience we shall relate these to several of the well-established age classifications as well as to some occasional nontechnically defined terms. The human *neonatal*, or *newborn*, period is defined as the first two weeks of life. The term *baby* refers to the first year of human life, which is by definition the first part of infancy. *Infancy* is commonly described as that time between the neonatal and the *childhood* period, which begins at approximately three years of age and lasts until the beginning of adolescence. It is during childhood that the truly social forms of play develop and become organized, and begin to operate as creative socializing forces.

PRESOCIAL PLAY

The age-mate affectional system is of superordinate importance for normal social and sexual development. In fact it now appears that the increasingly complex processes of play provide the means and motives for the development of the peer system. However, play does not spring spontaneously like an Athena from the head of Zeus, even though it is indeed a godlike gift. Developmentally and functionally, it progresses according to a definite maturational pattern. Play with inanimate objects precedes play with animate objects, so that presocial play by definition precedes social play of comparable complexity. For example, the tiny tot plays with his dangling mobile and musical bells before he is six months old, whereas the social-partner games of peek-a-boo and patty-cake develop some months later. The detailed development and interaction between presocial and social play in the rhesus monkey has been outlined by Harlow [1969]. The presocial play of the boisterous baby mixing his pudding and his cereal, rolling his peas across the tray, and pounding the spoons in clanging cacophony precedes any activity approximating social play. Months pass before these simple acts are transmuted into social behaviors, at which time children might be found exchanging sand trucks, rolling balls back and forth, or concomitantly pounding rhythmic patterns with kitchen lids and ladles.

We will briefly discuss three types of nonsocial or presocial play, the first of which is *exploration*. Simple examples of this are the child's playful exploration of his own body—touching toes, flapping ears, picking nose, and the inevitably disapproved act of masturbation. The second form is that of *parallel play*, in which two or more individuals play simultaneously and in close proximity, but without any apparent interaction. The final presocial play form is *instigative play*. In this case the activity initiated by one child serves as a model to another, who immediately accepts the challenge and attempts the chosen chore. The basic importance of presocial play and its effect on the child lies in the fulfillment of three functions: It trains the child in exploration of the environment; it weans him from the maternal figure as the sole security agent; it progressively prepares him to respond socially to the activities of peers. This progression is evidenced in the very nature of the three presocial play forms—exploration, parallel, and instigative.

EXPLORATION PLAY   Human children have presocial play capabilities unmatched by any other animal. The child drops his spoon on the floor and the mother replaces it religiously and relentlessly, again and again. Ultimately she may attach it to the highchair with a string and leave the retrieval to the infant's own devices. As mischievous behavior is reinforced, it gains in gaiety. From the child's point of view this is not social play, but asocial play. The mother is merely an inanimate object with marvelous cooperative capabilities. Other solitary play is that with a hobby horse or blocks, or looking at picturebooks and coloring with crayons. Solitaire may later represent a very nonsocial game of cards played on a table.

PARALLEL PLAY   Parallel play is playing beside others rather than with them. This type of play is commonly seen when two pretoddlers share the same play area but share absolutely nothing else. Each is fully content with himself and his toy and perhaps even the presence of his partner, but there is no desire to interplay. Thus the children engage in the manipulation of one or more objects without resorting to any social goal or social interactions, as shown in Fig. 2-1.

FIGURE 2–1   *Parallel play in human infants.*

FIGURE 2–2   *Monkey infants in close physical proximity.*

FIGURE 2–3   *Parallel play in macaque infants.*

INSTIGATIVE PLAY   Antecedent to the stage of true social play, but subsequent to the stage of parallel play, are the activities of a child which serve as initiators or motivators for other children but do not lead to social interaction *per se*. Such evoked activities may be conceived of as social sequential responses. They are socially motivated but are not necessarily socially achieved. Examples of such behavior include follow-the-leader, imitation, mimicking, the copy-cat routine, or monkey-see monkey-do.

MACAQUE PRESOCIAL PLAY   Play ranging from solitary to sequential appears in monkeys early in the first month. The members of a group of infant monkeys, as shown in Fig. 2-2, tend to remain in close physical proximity to one another, and when one member physically detaches himself, the others may visually orient to him and follow in sequence. Transient and relatively aimless oral, manual, and total bodily contact may take place without any social feedback being expected or received.

A crude type of parallel play is shown in Fig. 2-3, with early infant monkeys exhibiting parallel activities toward a single object. Whether or not monkeys engage in human baby-type parallel play is an open question. As the monkeys' motor capabilities mature during the first month, patterns of sequential play are commonly seen. A monkey climbs a wire-mesh ramp, and a second monkey follows; a monkey walks across a rod high above the floor, and one, two, or three monkeys may follow in sequence. One monkey finally swings free from a ramp to a flying ring, and other observing monkeys then approach the flying ring in sequence.

SOCIAL PLAY

As presocial play wanes, more complex and more socially demanding forms of play appear, and these comprise the multitudinous categories of social play. There has been a spate of studies on social play, but most of the studies of human children have been observational studies in the home or nursery school or questionnaire researches on parents. Nursery school studies usually stop when the child is five, when play is literally in its early infancy. Homes and nursery schools greatly restrict many forms of play, particularly those judged by homemakers to be antisocial or inimical to orderly operation of clean kitchens or pretty parlors. As for the play questionnaires, most of the items appear to have been formulated by men whose memories of childhood play suffered severely from critical degeneration, retrospective falsification, or Freudian inhibition.

Social play may properly be divided into three major forms—free play, creative play, and formal play. Free play can be conducted without recourse to formal rules and may be physically vigorous and even violent, or sedately satiating. Rough-and-tumble play, cops and robbers, and chase, as well as playing house, school, and store exemplify free-play activities. Either physical or cognitive factors may predominate, but the outcome of the activity is not predetermined, and there is no prescribed process. Formal play, as the name implies, is conducted within the limits of prescribed or proscribed rules. In games such as hide-and-go-seek, London Bridge, and drop the hankie, the rules and regulations have been almost identical for hundreds of years. Formal play is doubtless limited to man, since language is almost an essential adjunct. Most other animals would inevitably find formal play a dull and deadly drag. Creative play differs from free play and formal play in many ways. One of these is that creative play is not unique to any fixed ontogenetic phase, but can be found throughout a series of ontogenetic stages. Creative play may be constructive and as simple as using a dishtowel as a doll blanket, or it may be as complex as the creative contributions of Shakespeare or Einstein.

FREE PLAY   In the conventional depictions free play is exemplified by games of catch, running, and romping. Certainly this is the most basic type of free play. However, there are free-play forms which transcend the physical. Beyond the physical there are types of free play which are primarily characterized by cognition. A prototype is invention. Finally, there are free-play forms in which cognition alone is not enough, but must be blended or modulated by affective tone. Within this category lie most works of art, such as painting and sculpture, musical composition, and writing.

Of all the types, physical free play is the easiest for the child, and the most disturbing for the parent. Despite its simplicity, this form of free play plays a predominant role in the socialization process, for it is here that social ordering and even social roles develop, that the rules of social intercourse are shaped, and eventually, that the control of immediate demand and aggression is established. Unfortunately there is almost no free-play research with human children—and there is some question as to whether most of the information we do have represents data gathering or woolgathering. We cannot help but think that much of

39

FIGURE 2–4   *Rough-and-tumble play.*

the observational data are poisonous parodies of physical free play's better and best behaviors. In spite of the lack of human data, the social and psychological impact of free play is amply documented by meticulous macaque monkey studies. Fortunately there is also one effective human study modeled after this monkey material [Jones, 1967]. For indeed, the best way to find whether monkey behavior can be generalized to man is to discover that a parallel already exists.

No other single form of play is more important to basic socialization in the monkey than physical free play. Frequency and finesse of social interaction in such play determines social status. Such contact play also shapes and shifts social roles.

Roughhousing monkeys wrestle and roll and sham bite, but no one ever gets hurt and no one cries out in pain (see Fig. 2-4). While the participants energize, of course, the parents agonize. The development of contact or roughhouse play in rhesus monkeys is shown in Fig. 2-5. It is obvious that this form of free play increases in frequency and force until the age of one year and then remains at a plateau for a considerable period of time. Since the entire laboratory staff was unsophisticated concerning sex differences when these experiments were planned, the enormous difference between the male and female babies in rough-and-tumble play was totally unexpected. The data presented in Fig. 2-5 give strong presumptive evidence that males and females are innately different in their approaches to play. This significant difference between the sexes is doubtless the primary biological basis for the subsequent cultural shaping of appropriate sex roles.

In addition to rough-and-tumble play, experimenters have observed and recorded a form of noncontact play or approach-withdrawal play. In this play pattern the infant monkeys chase each other back and forth, up and down, and round and round. Frequently the role of the chaser and the chased alternates, and occasionally, but rarely, brief body contact is made. The social importance of this play cannot be denied. It appears, for example, that if young female monkeys are not chased early, they remain chaste forever. Noncontact play matures a little later than contact play, but typically the two play forms intertwine for a long developmental period. Actually, there is now a considerable body of

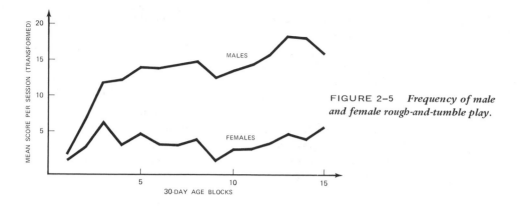

FIGURE 2–5 *Frequency of male and female rough-and-tumble play.*

data showing that noncontact chasing play is probably predominantly feminine. It certainly would not be surprising to discover innate sexual differentiation in play patterns. Such differences may uncover a secret to sexual segregation. They would be one of the unconditioned responses that are basic to learned social and cultural behaviors.

The British child psychologist Jones [1967], who trained with Tinbergen as an ethologist, tested the Harlow model of monkey rough-and-tumble play on human infants through free observation of infants five years of age and under in three nursery schools. Jones found that these children spent most of their time with age mates, and the teacher was relegated to the role of a distant agent disbursing basic comfort, care, and trust as a positive reinforcing mechanism for the expression of free play. Jones has concluded that "almost identical patterns [to the monkey pattern] of play occur, and are clearly definable, in human children." As an ethologist he describes this kind of free play in children and, with objective propriety, outlines the components of human responses as differing from those in the monkey, which, of course, they do; man is more than a monkey captured by culture. Human components include jumping up and down, open-handed lashing, laughing, and making mischievous faces; these are obviously human fixed-action-patterns evoked by social stimulation.

Human rough-and-tumble play looks hostile to the outsider, adult or peer, and it is agonizing to mothers, but it is an activity of gaiety and charm for the participants. Jones found rough-and-tumble play in human children as early as eighteen months and as late as three to five years. He points out that this primitive free play develops rather sharply into formalized games such as tag and cowboys and Indians, in which the same motor patterns apply but rules and verbal explanations have been added [Jones, 1967]. This is an obvious point at which monkey children and human children part company; monkeys do not play by verbalized rules.

Jones interprets the rough-and-tumble play in children as non-aggressive, although it is vigorous and must be basic to social-dominance role formation. Limited data suggest that human rough-and-tumble play, like monkey rough-and-tumble play, is much more masculine than

41

feminine, and that this difference is not culturally determined. Such sex differences in free play have wide implications for social and sexual development. As Jones points out, this infantile vigorous but joyous play pattern is "such conspicuous behavior that one could not have long disregarded it and its effects in an ethological study of children." It is quite obvious that rough-and-tumble play could lead to aggression, when aggression matures. However, it is unlikely that real concerted aggression has developed to any large extent in the four-year-old human child, reports from nursery schools notwithstanding. In retrospect, Jones' study serves as a clear reminder that human investigations, which are subject to bias, can overlook the obvious.

CREATIVE PLAY   Creative play begins when the child becomes the master, the creature becomes the creator, and all reality becomes subservient to the child's whimsical whirlwind of wishes. In creative play the child uses as raw materials things which to adults are finished products. He purifies and personalizes such objects to serve him on so unique a basis that neither trademark nor patent is necessary. This stage of creative play may be broadly defined as the use of objects for purposes other than their original ones. The child who constructs a staircase for dolls from blocks is enjoying creative play; so is the snow-soaked boy who builds a snow-packed dam in the gutter outside the house. Elements of mastery and self-actualization are seen in these creative constructions as the stairs ascend and the dam rises to the full limits of the impending gravity or turbulent torrent. What Piaget [1967] identifies as symbolic play and defines as the apogee of all children's play is, in this sense, an early aspect of creative play.

Unlike most free play, creative play is in no sense restricted to the activities of children. For example, an enormous amount of artistic effort—painting, sculpture, prose, and poetry—falls within this category. Of course motives other than pure entertainment might simultaneously be involved. Such motives might be economic, narcissistic, and even romantic. Attempts to identify and measure creativity [Roe, 1952; Taylor, 1964] have not been outstandingly successful, in spite of the fact that these attempts were made by people who are themselves recognized as creative. However, there seems to be relatively strong agreement on the characteristics of the creative individual. Less can be said of the available information regarding creative play. Psychologists have done virtually nothing to examine the effects of creative play. Is it a prerequisite for general creativity? Does it lead to withdrawal from reality or to a deeper involvement with reality? What accounts for the apparent decrease in creative-play activity during adolescence? How, when, and under what circumstances is it revived in adulthood? These are only a few of the questions which should concern the child psychologist and educational psychologist.

FORMAL PLAY   The moment a person surrenders his early creative-play position and possessions, he becomes a puppet. His moves are restricted and his movements restrained. Formal play is the dictatorship

42

of recreation. One is given rules, bylaws, and a credit system, and for each move or alternative the consequences are predetermined. This fatalistic fun is enjoyed by those who play ring around the rosie or drop the hankie, basketball or baseball, pinochle or poker. From a developmental standpoint there appears to be irony in this ritualistic recreation. As physical and mental independence increase, we tend to favor forms of play which have increased restrictions. Our recreational preoccupations are transformed from artistic adventures to scientific solutions.

While many of the important effects of free play are provided almost exclusively through animal research, no companionate body of information exists for formal play. Whether the human effects of formal play are restrictive or rejuvenating is a decision based at best on objective observation and at worst on sheer speculation. A problem of paramount importance in this connection is the influence of formal play on the apparently spontaneous efforts of the younger child. Does formal play stifle or stimulate, crush or create, integrate or segregate? These seem to be questions of concern for many but research topics for none. Even the parent-teacher associations recognize that creativity must not be stifled. They worry about the effect of classroom regimentation and formalization on spontaneity. Unfortunately educational researchers have failed to translate this clear concern into meaningful research. It may be that researchers are so preoccupied with work in a limited sense that they take pride in not being occupied with play. Whatever the reason, the child psychologist's contribution to the theory of human play warrants a modification of Churchill's eulogy to the British airmen: seldom have so few done so little for so many.

THE FUNCTIONS OF PEER PLAY

The various forms of presocial and social play may appear to defy their definitions, but this is a defiance in fashion rather than fact. Thus a child captured in a playpen experiences social constraint, but the awareness of an amused audience may make his play activity far more social than presocial. Similarly, the participant in an art contest may retreat into self-designed solitude and work unknown and unrecognized, but with the persisting presence of his imaginary audience as his primary motivation.

Exploration is not restricted to presocial play, although it originates at that stage. A lone child may visually examine and investigate a large culvert or tunnel but not dare to enter to culminate the acts of exploration. However, let him be joined by his gang of peers, and they will scurry to and fro through the tunnel without a tremor of fear. These examples of compounding and overlapping play forms not only give rise to complex play patterns, but also terminate in goals which are quite different from those achieved by any one single play form. For example, our monkey data show that presocial exploration is enormously

43

inhibited by the slightest fear, and this also appears to be the case with children. Social exploration, however, obliterates fear to a vast extent and encourages unrestrained exploration, which further reduces the fearsome aspect of the child's wondrous new world.

Friendship often begins in mutual exploration of some common object, task, or interest, but the fruit of such activities frequently includes subjective as well as objective discoveries. Thus the intertwining of presocial warp and social woof creates the color, complexity, and constructiveness of the child's play patterns. It is these complex patterns that facilitate the productivity, durability, and enrichment that characterize peer harmony. In this respect our classification of play forms fails to suggest the full fruitfulness of age-mate play functions. Classification is a hard task, and it forces us to digitalize that which in fact is a continuum. It is the haven of the professor and the hell of the student, who is forced to memorize as facts distinctions which at times are fiction. Nevertheless, classification does serve as a scanning device, permitting us to see some order in a situation otherwise bewildering in its perplexity.

By the same token, the paeans of peace which a mother sings for her child are achieved far less by her efforts than by the interactions of age mates. The primary basis of aggression control is the formation of strong, generalized bonds of peer love or affection. Fear may be thwarted by mothers, but aggression is controlled only by age mates. All primates, monkeys and men alike, are born with aggressive potential, but aggression itself is a relatively later-maturing variable. It is obvious that a one-year-old suffers from fear and is terrified by maternal separation, but the child neither knows nor can express aggression at this tender age. In fact he may long be hampered from expressing any aggression, for a child quickly learns that it is culturally unacceptable for him to be aggressive toward his parents, and frequently these are his only available associates.

This lack of aggression targets accounts in part for the fact that "evil emotion" culminates during the age-mate stage, long after peer affection and love have developed. It is the antecedent age-mate love that holds the fury of aggression within acceptable bounds for in-group associates. Thus one of the primary functions of peer play is the discovery and utilization of social and cultural patterns. Play acquaints the child with the existence of social rules and regulations and their positive as well as negative consequences. This may be looked upon as a rediscovery of the reality principle—not the individual-reality principle, which the infant discovers through interaction with its mother, but the social-reality principle discovered through interaction with his peers. Familiarity with the established social and cultural patterns offers the child the reward of social acceptance, the freedom to engage in play of challenging complexity, and guardianship against social failure and rejection.

With the discovery of this social-reality principle the child acquires new freedoms as well as new restrictions. For example, the fears and freedoms of a child newly initiated into a peer group are quite different from those he harbored prior to membership. The moral support supplied by the group diminishes the fear of real objective danger and intensifies

44

the fear of age-mate rejection. Thus it is not uncommon to find the child more self-conscious, although he gives the appearance of being far more overtly extroverted.

One of the primary socializing functions of age-mate or peer affection is the opportunity for the formation of personal love bonds. Most commonly these friendships are between members of the same sex because sexual differentiation of interests and availability has been established. However, friendship bonds are also established between opposite-sex members, and these friendships may be maintained for long periods of time at purely platonic levels. At the baboon level friendship may persist between males that are the dominant members of the troop. Neither age nor physical strength is a major variable in the establishment of these bonds. In fact friendship can be so overpowering that these, as well as such variables as physical attractiveness, intellectual ability, and even sex, may be subservient to the mysterious, unique, unidentifiable force that forms a friendship. This is one of the powers of peer passion—it cuts across age-mate relations to prepare the individual for adjustment to the inevitable.

HETEROSEXUAL LOVE

While the path to passion is paved with play, the heterosexual passion play is physically, behaviorally, and culturally distinct from the adolescent and adult forms of peer friendships. The heterosexual affectional system typically emerges at puberty, reaches full expression by late adolescence, and operates throughout most of adult life. We are not particularly concerned about the function of this system, since this is a matter of common—often extremely common—knowledge. Our interest in the heterosexual system centers on its developmental analysis—the variables which facilitate the development of this system into its culturally standardized form, and the variables which distort, disrupt, and destroy it.

Heterosexual affectional relations develop in all primates through three relatively separable subsystems—a sequence of postural potentialities, elicited by external stimuli and leading to the complex interbody positioning which adult coital behavior requires; a flow of gonadal gifts which indirectly and directly facilitate heterosexual interactions beginning at puberty; and an affective model built during the infant and age-mate love systems and applied, sometimes forever, after puberty. In other words, the heterosexual system appears with the development of mechanical sex, secretory sex, and romantic sex.

The postural problems in the development of the mechanical subsystem range from the relatively discrete reflexes, such as penile and clitoral erection and pelvic thrusting, to the complete coital combinations performed by adults. These responses are determined in part by anatomical structure, in part by basic sex-differentiating responses, and in part by environmental conditioning which shapes our destinies.

45

A great maturational gulf separates the appearance of mechanical sex from secretory sex. Components of mechanical sex appear from birth on, while secretory sex is the signal of puberty. Secretory sex operates through the action of gonadal hormones upon discrete sensory and motor elements of sexual behavior. The appearance of this subsystem in the female is marked by menarche and follows the cyclic pattern of hormonal fluctuations during the monthly menstrual periods. Secretory sex is acyclic in the male and is marked by the appearance of the ejaculatory reflex. The delayed timing of gonadal glory allows the two other heterosexual subsystems, the mechanical and the romantic, which emerge more gradually, to reach the development necessary for complete heterosexual relationships. Another difference between mechanical and secretory sex is the path through which each makes its influence felt. Mechanical sex is elicited by external stimulation, while gonadal hormones exert an internal influence on sexual behavior.

A third difference between these two subsystems is that the advent of mechanical sex is influenced by cultural variables, since its open expression can be either delayed or advanced by cultural disapproval or acceptance, or any other form of learned inhibition or reinforcement which individuals experience. Secretory sex is resistant to these influences since it operates in secret through internal physiological mechanisms which are insulated against cultural condoning or condemning. These two subsystems also differ in that social isolation early in life exerts profound negative influences on sexual mechanical expression, but learning leaves hormonal maturation and sexual excitement undisturbed. Nevertheless the failure to develop affectional bonds early in life has a profound impact on the naked realities of sexual performance. The power of love to mask love's labors is most evident in the emotional depression which may result when strong affectional bonds are disrupted. Sex secretions may create sex sensations, but it is social sensitivity that produces sensational sex.

Neither postural potentials nor gonadal glory are totally adequate to express the full range of heterosexual relationships in primates. They are particularly unproductive in providing the variables underlying romantic or idyllic ecstasies in humans. Beyond these two subsystems, there is a third one, which we commonly refer to in humans as romantic love and in monkeys as transient heterosexual attachments or preferences. This subsystem includes an emotional component of affection and the behavioral patterns defining masculine and feminine sociosexual roles. The affective component goes beyond the emotions of mechanical-secretory sex, and identifies the basic nature of the entire system. However, even this affection does not operate in a vacuum, and the context of male and female gender roles is required before romantic love can have any content or consequences. Masculine and feminine gender roles transcend postural differences during coital conquest and pervade courtship, companionship, child care, and community commitments.

Like the mechanical heterosexual subsystem, romantic love has

46

innate and developmental roofs in the preceding infant-love and peer-love systems. The basic trust gained from maternal love, the elaboration of contact acceptability, and the pleasures of propinquity acquired in age-mate play operate in the transition of peer affection to heterosexual love. The sex-differentiating behavioral patterns comprising adult male and female roles and social status also develop gradually from infancy onward. Like mechanical sex, the arousal of the affective component and the display of sex-appropriate behavior are elicited at times by external stimulation. For example, Paul Newman and Natalie Wood have acquired fame and fortune through their celluloidal capacity to create affective arousal which leaves females gasping and males groping. Similarly, exaggerated sex-role behavior prevails predominantly among the impressionable.

The third heterosexual subsystem is crucial for the future production of both monkeys and humans. However, it is vastly more complex in man than in monkeys because of the influence of a wide range of cultural variables. Through the increased capacity for abstract conceptual thought provided by language, human social life has complicated tremendously the basic primate patterns of social organization, and these cultural variables exert a pervasive influence on romantic attachments through children's learning about the culture's model for heterosexual relations. This model varies widely in different cultures, from promiscuity to monotony. Whatever the model, however, children accept it and expect to achieve it long before it acquires any functional significance for them. Romantic sex is also far more vulnerable than mechanical sex to early social isolation, since primates, especially male primates, who have not experienced love early will never learn to love, and social behavioral development cannot occur at a distance.

The romantic affectional subsystem also shares some of the basic features of the secretory subsystem. While it has early developmental origins, it is elaborated by maturational factors and by heterosexual learning throughout the growth period for secretory sex and reaches peak development at middle to late adolescence in those who are fortunate and not too foolish. This adolescent period of yearning and learning results in the assumption of adult roles in social groups of monkeys and men. It begins when promotion to puberty necessitates a redefinition of roles. Thus adolescents face the task of becoming familiar, through specific learning, with new modes of interaction with individual members of the opposite sex. Unchaperoned adolescents are almost universally successful in this learning. Parents of females in many cultures forget their own learning period and worry that their children are not facing the future cheek to cheek, that their learning is far too specific, or that practice sessions may not be limited to a single individual. Psychologists can assure such parents that their children are employing sound learning principles, including the partial reinforcement of unprepared parentage as a consequence of lack of precaution. Romantic love involves internal factors as well as external factors, and like secretory sex, these are mediated by the brain, although it is to be hoped,

47

by less primitive centers of the brain. These internal factors take the form of strong motives to be loved, and to be loved by a member of the opposite sex, acting as that sex should act. The need for close social bonds, the formation of gender identification as male or female, along with acceptance and adoption of sex roles and of the cultural reproductive model, are relatively permanent characteristics of human adolescents which operate with the strength of basic moral obligations.

Each of these three subsystems of the heterosexual affectional system provides independent contributions and can even operate to a limited degree in the complete absence of the others. Nevertheless, they normally operate together in adult male-female relationships. While there are doubtless many complex interactions among the three subsystems, our knowledge is relatively certain for only a few of these. For example, it is well-known that gonadal hormones act to lower sexual sensory thresholds, intensify sexual stimulation, and decrease the period of sexual arousal required to satiate coital responses, especially in males. A powerful interaction of the third subsystem of romantic love on the first is the total elimination of mechanical sexual responses when the capacity for male-female affection has been destroyed or distorted by inadequate or abnormal affectional opportunities early in life. This is not a regular occurrence in modern child-oriented cultures. It is, however, by no means unknown in humans who have been raised in institutions under impoverished social stimulation, or who have been raised by adults with little maternal merit. The total and often irreversible social crippling displayed by these few affectionless individuals provides striking evidence of the overwhelming importance of the affectional subsystem in normal heterosexual relationships. The importance of affection to sexual relationships is also revealed in the ease with which problems in the mechanical and secretory subsystems can be solved in the context of warm, intimate affection between a heterosexual pair. Problems such as premature ejaculation in the male, and vaginismus (vaginal contractions which totally prevent penile intromission) can be corrected within weeks by means of informed sexual procedures and a generous supply of remedial romance [Masters and Johnson, 1970].

This outline of the nature and variables of the heterosexual subsystems, their developmental trends, and their interactions, characterizes this affectional system in man and in monkey. The empirical basis for the outline comes in large part from experimental studies of heterosexual behavior in monkeys under controlled rearing conditions. These findings are described below to illustrate the bare facts of life. Similar studies of humans are, and in some cases must remain, nonexistent. Where partial human data exist, they in no case change this basic outline, although the human behavior is vastly more complex and these complexities will be considered. Abnormal heterosexual behavior often exists more in the minds of the created than the creator. Nevertheless, a separate section on this topic is included in order to reach the larger and more affluent of these two audiences.

48

The monkey neonate and the maturationally equivalent human infant possess two innate components of mechanical sex, even though their likeness to love is purely coincidental. These are penile erections, which are more prominent than clitoral erections in the female, and pelvic thrusting, which is more common but no more prominent in infant males. The normal and natural stimulus for infantile pelvic thrusting is a warm, soft body. The mother first provides these conditions, and any human mother who disclaims knowledge of this pelvic thrusting is either repressed or a bad maternal bet. Monkey research and competent psychiatric research leave little doubt that sexual reflexes involved in adult sexual behavior are present as soon as physical maturation will permit their display.

The next steps in the development of monkey mechanical sex achieve approximations to the essential components in adult male-female connubial combinations (see Fig. 2-6). One of the essential components of this adult pattern is the anatomical requirement for ventral-dorsal positioning of the male against the female. Another essential feature is that the female must support the male, at least in monkey societies, by standing on all fours and rigidly elevating the hindquarters. A third requirement is that the male must achieve genital propinquity by mounting and grasping the female's hindquarters with hands and feet and must achieve genital interaction by repetitive pelvic thrusting. These three features exhaust the postural possibilities for monkeys, although sex can be more stimulating to simple simians if the female's forelimbs are flexed, and if the female frequently reaches behind and looks toward the male, possibly in awe and admiration.

The first two components for this species-specific pattern, ventral-dorsal positioning and rigidly elevated hindquarters, grow out of three discrete responses—threat, passivity, and rigidity—during infant play. These three responses appear with regular frequency in monkeys by sixty days of age, soon after play begins—and they occur with statistically significant difference in frequency in males and females. As the data

FIGURE 2-6   *Basic adult sexual postures.*

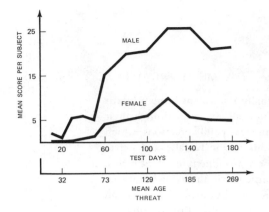

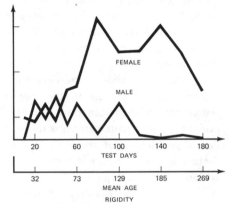

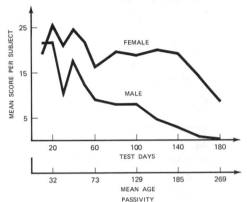

FIGURE 2–7 *Development of sexual postural responses.*

plotted in Fig. 2-7 demonstrate, threatening is a masculine prerogative, while passivity and rigidity are female prerogatives. Rigidity is the infantile precursor of the adult female sexual-presentation and sexual-support postures. When it is displayed by the female infant, after the male infant threatens and the female withdraws, stops, and waits (passivity), the ventral-dorsal situation is created by an assertive male approaching his diminutive playmate. But, while females presexually play by standing steadfastly, males presexually play by gripping and grasping. Figure 2-8 reveals that the male presexual pattern of hindquarter grasping, later followed by thrusting against a new, soft, and warm surface, is soon added as the third basic sexual component growing out of peer play. With all three components present, the approximation to the adult pattern is achieved with complete innocence (Fig. 2-9). At this point the sexual positioning pattern of adults is only one learning step away. Grasping and thrusting increases with age, as shown in Fig. 2-10, and a similar increase occurs in the complementary female rigidity reaction. This natural developmental sequence ensures that males and females will be at the right place at the right time when organismic opportunity occurs at puberty.

The appearance as early as sixty days of responses basic to adult patterns suggests that these are innate responses to specific external stimuli. Furthermore, their increase and refinement with age occur in the absence of reinforcement by sexual satisfaction or cultural comment.

50

FIGURE 2–8   *Basic presexual position.*   FIGURE 2–9   *The first presexual step.*

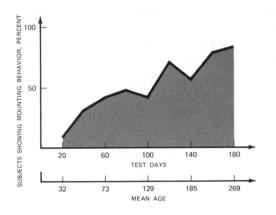

FIGURE 2–10   *Development of mounting behavior by males.*

The data on overt total-body sexuality in human infants are hidden in horror. It is reputed that such behaviors are seen in some more relaxed cultural communities, such as Indonesia and Appalachia, but we can describe no systematic developmental data. Obviously the particular postural potentialities would be different in form and related to the time and type of human play. The data collected by researchers in our society using survey questionnaires loaded for lust are devoid of broader developmental patterns and are both disappointing and depressing [Ramsey, 1943; Elias and Gebhard, 1969]. Data collected in this manner measure the end product of a hopelessly confused complex array of biological potentials, environmental and cultural conditioning, and the sex differences along these dimensions. However this complex array of variables operates, we discover that infantile self-exploration is present in humans, that limited and basically aimless sex play occurs in childhood, and that human males not only display more of everything than females, but also meander to maturity through masturbatory marvels. Males in our society have obviously written their own sexual developmental script by hand.

The details described for the development of mechanical sex in monkeys do not imply that mechanical sex is critical to the development of heterosexual affection. These responses do, however, illustrate that relatively complex behaviors can have a biological basis and emerge through mechanisms other than cultural ones. Moreover, there is no dis-

51

continuity from unlearned to learned behaviors, and complex capacities may well depend on a less than fortuitous combination of both. These responses also establish that there are innate sex differentiation mechanisms related to natural and normal social behaviors in primates, which undoubtedly operate in many aspects of social development in man as well.

THE SECRETORY SUBSYSTEM

Human females experience first menstrual bleeding at an average age of thirteen, which is the maturational equivalent of thirty months in rhesus monkey females. Using years when comparable data were available, McCammon [1965] reports that the age at menarche in humans did not change from 1930 to 1965, and anecdotal accounts of ancient China also placed menarche at thirteen years. Cultural variations and associated environmental changes apparently exert little or no influence on this subsystem in females. Data on equivalent pubertal age in human males is unavailable, but nonsexual measures of pubescence suggest that, like rhesus males, they lag significantly behind females.

Gonadal hormones in adulthood could conceivably influence the heterosexual affectional system directly or indirectly. A direct influence, for example, might affect the intensity or patterning of motives comprising the heterosexual systems in males or females. Direct and total control of human sexuality by secretory factors was a commonly accepted belief before the twentieth century—that is, before Freud and before the scientific study of sexual behavior. People then believed that deprivation of sexual activity in adult males created an irrepressible physical urge which could be relieved only through overt sexual expression. This belief functioned well for the management of guilt in men with unmanageable sexual appetites, but it did not survive empirical scientific digestion.

Gonadal hormones are not necessary for adult sexual expression in higher primates. The amazing amorous achievements of human and monkey males who are denied endocrine excitement, either by design or by disaster, demonstrate that the primary motivation for heterosexual relationships originates above the shoulders. Similarly, confessions of conjugal complicity by regularly menstruating and sexually active females reveal no systematic relationships between their secretory fluctuations and their sexual fantasies, frequencies, or fulfillments. However, female activity is so confounded by fear of conception that we may be at least two generations away from the necessary contraceptive calm. A more likely role for sex hormones in human sexuality is their influence in enhancing the affective impact of erotic stimulation. In this way hormonal action would contribute to learning that overt sexual responses produce positive affect, which could lead to the establishment of a pleasure seeking sexual motive. Once established, this hedonistic principle could operate alone even if hormonal support were withdrawn.

Whatever the direct contributions of the secretory subsystem to heterosexual behavior may be, it probably has far more important in-

52

direct effects in humans through the cultural and personal significance of the secondary sexual characteristics, which develop rapidly at puberty. The mediation of these effects through cultural influences can be demonstrated with a simple example of the significance of early and late maturation for boys in our society. The physical signs of puberty are taken as the mark of a dramatic leap toward emotional, cognitive, and social maturity. Consequently physically accelerated boys are treated as more mature by both adults and peers. They are rated as more popular and socially desirable and are given increased social status. Late-maturing boys, who do not receive this enhanced respect, become anxious about their status and adopt compensatory attention-seeking behavior, which in turn adds to their adjustment problems [Jones, 1954]. Cultural influence in mediating the transition from preadolescence to adolescence is often exerted through the rituals and badges which symbolize cultural recognition of new reproductive status and new social roles. The use of adolescent badges is probably culturally universal, although it is unknown in primates other than man. Beauty, however, is culturally arbitrary, and cultural conditioning determines whether unbelievable feminity and beauty are to be achieved with lipstick or labial lacerations.

Physical maturation and cultural recognition typically operate together to usher in the period of heterosexual potential, and their effects cannot ordinarily be separated for scientific study. However, the extraordinary confusion which occurred in the culture of Pukapuka demonstrates that cultural conditioning plays a powerful role in terminating the prepubertal period of sexual latency.

The island of Pukapuka is a tiny coral enclosure in the balmy paradise of the near South Pacific. For countless years fish teemed in the enclosing coral arms and the palm trees proudly presented their carbohydrate complements. The inhabitants were healthy and happy and such dominant hierarchy as there was existed in fashion and fiction only. In Pukapuka it was customary and conventional for the children to wear no clothes, since clothing had a special significance for postpubertal preparation. On warm tropical nights, when the full moon climbed heavenward over the bright blue waves, the adolescent boys would gather together and walk aimlessly eastward, and the adolescent girls, with their beauty for the first time clothed in bikinis, would gather together and walk aimlessly westward. By some unbelievable coincidence the two groups would meet and then disband as heterosexual couples. This was known as postpubertal preparation in Pukapuka.

All continued contentedly until the British missionaries arrived and were shocked to see unclad prepubertal children. They quickly resolved the problem by edict. All prepubertal boys and girls henceforth wore clothing. To the simple Pukapukan mind this had but one meaning—on warm tropical nights, when the full moon climbed heavenward over the bright blue waves, the prepubertal boys would gather together and walk aimlessly eastward, and the prepubertal girls would gather together and walk aimlessly westward. . . .

53

The essential characteristic of the romantic subsystem is an affectional bond between the two members of the heterosexual dyad. This affectional bond probably operates in some degree in every consenting heterosexual relationship which primates form. Anthropomorphic and cultural bias teaches us to identify this romantic subsystem with the ecstatic happiness of a cherished monogamous relationship. This same cultural bias categorizes deviations from this model as being devoid of romance, views cultures with less restrictive arrangements as primitive or semi-civilized, and considers heterosexual behavior in nonhuman primates as mildly amusing examples of sexual depravity. However, comparative heterosexual analysis suggests that there may be less significance in the heterosexual variation among primates than in the variation between primates and more primitive mammals and vertebrates [Beach, 1969].

In most primitive animals heterosexual and other social behavior is under strict control of environmental variables such as temperature, hormonal variables, and external stimuli, such as the plumage of birds, which does no more than identify species and sex. These variables alone produce fortuitous and indiscriminate heterosexual dyads in primitive mammals and vertebrates for the period required for fertilization. A few species of birds are capable of individual recognition and form longer lasting relationships. These relationships, however, usually last no longer than the breeding season. They entail little behavioral variation in individual pairs from the species-specific pattern and are governed by the principle of impersonal, affectionless sex.

In contrast, all primates which have been systematically studied depart significantly from indiscriminate promiscuity. Their relationships are characterized by selectivity in pair formation, determined by affectional compatibility of individual pairs. This new principle of affectionate sex sets primates apart from more primitive animals and is a common element in all primate heterosexual behavior. Of course one rodent, the American beaver, some of the carnivores, most notably wolves, and at least one species of deer also display this selectivity, and some even form monogamous relationships for life. The scarcity of data on monogamous mammals does not imply that monogamy is the misfortune solely of primates. In fact, with the exception of the darling deer, our stereotypes of some animals are questionable; not all rodents are rats, and the wolf whistle is, in the wolf, a welcome only to faithfulness forever.

These comparative data suggest that the variations in human, ape, and monkey heterosexual relationships may be viewed as differences in the degree of affectional influence on the time and behavioral content devoted to the courtship phase, the conquest phase, and the consequence phase of the heterosexual bond. From this standpoint we can see similarities between the 15-minute march to conquest and the prolonged courtship and conquest involved in matrimonial morality. Thorough or limited selectivity is professed in both cases, and the affectional bridge leads to both bedrooms. In monkeys and apes, under natural conditions heterosexual attachments are usually limited to consort pairs which last for hours or days. Such sexual arrangements might be described as brittle monogamy. The presence of a basic primate pattern for pair formation does not imply that there are not vast species and cultural gaps underlying heterosexual variations. However, affection as the foundation for heterosexual relationships in primates fosters the interpersonal propinquity in which learning, especially learned cultural complexities in man, may produce almost limitless variation.

Research has shown that humans and monkeys possess the same behavioral mechanisms which provide the transition from the preceding infant-love and peer-love systems to the romantic subsystem of the heterosexual affectional system. These transitional mechanisms lead to the basic requirements for the formation and functioning of the heterosexual affectional system. These basic requirements are heterosexual trust, acceptance of heterosexual contact, behavioral sex-role differentiation, and social motivation for sheer physical proximity. Although these four ingredients are necessary for all the major features of heterosexual relationships to appear, basic trust and acceptance of heterosexual contact are an absolute requirement for coital captivity. These four factors can also create the opportunity for additional learning at puberty, which may complement and elaborate heterosexual relationships. For example, as we have discussed, hedonistically motivated sexuality may develop as mechanical and secretory sex become fully functional.

While these four requirements adequately describe the common elements in all primate heterosexual behavior, in humans additional transitional mechanisms, as yet unknown, are required to explain the enormous cultural and individual complexities which the learning and language capabilities of humans provide. These learned variables probably operate through the exceedingly complex gender roles which human males and females acquire. Gender-role development in humans involves cognitive identification, acceptance, and adoption of the appropriate role. These roles are determined by anatomic variables, innate behavioral variables, and a host of cultural variables defining the prescribed heterosexual model for adults.

Unmistakable developmental reflections of all four transition mechanisms are seen in the peer-love system, and their extension to the heterosexual system requires only brief description here. Heterosexual trust is seen as the management of fear during male-female interactions. This fear potential is different from earlier fear, since it is created by a physical and emotional intimacy which is more intense than that of any preceding contact. Fear in the heterosexual relationship can be a monumental factor, especially in males, because of the sheer physical exposure. The most vulnerable surfaces of the body are openly exposed in the compromising postures required in adult sexual relations. In addition, sex occurs with an adult, and sometimes around other adults, in whom aggression is fully developed, along with the physical size and strength to inflict deadly damage. Fear generated by the need for physical safety is far more important in monkeys than in man, since human

55

culture usually provides for physical privacy in such situations. On the other hand, emotional vulnerability, which arises from open emotional intimacy, is probably more characteristic of human than monkey heterosexual relationships when the cultural model includes strong affective involvement. Greater emotional intimacy enriches human affection; it also covers psychiatric couches with exploited victims of love's betrayal. An analogous deep depression in chimpanzees—and in a few cases in monkeys—is produced only by total physical separation of pairs which had initially became partners in love through artificial long-term propinquity in captivity.

Acceptance of heterosexual contact is merely the sequel to the acceptance of nonspecific contact, an early mechanism in the transition from infant love to peer love. Humans and monkeys are similar in their capacity to develop an abundant amount of heterosexual contact acceptance from their earlier presexual play parties. This foundation of posterior knowledge will lead easily and naturally to learning the adult sexual knowledge necessary for creating babies or committing adultery. Monkeys in all social groups achieve this knowledge as infants and as adults without the aid of instruction or culture. Humans in less restrictive cultures also achieve this knowledge playfully. In restrictive cultures, where this path of presexual play to adult sexuality is prohibited, children must rely solely on their distinctly human capabilities to conceptualize and imagine heterosexual content, and they typically reach adulthood with a pattern of marginal expertise and expectations based on myth. Humans also differ from monkeys in their capacity to conceptualize their creator and their conceptions, which may add a dimension of guilt and fear to sex that is unknown in monkeys. However, conceptual and cultural variables seldom substitute completely for the absence of peer presexuality, with the result that human adolescents are typically faced with the task of developing acceptance and knowledge of heterosexual contact in the face of total ignorance or crippling inhibitions. Fortunately, through the remedial wonders which an intimate affectional bond can create, such knowledge and acceptance can be discovered and developed by human adults.

The behavioral sex-role component of the romantic subsystem includes the development of all those characteristics which we associate with masculinity and femininity. Development of gender roles is a vastly complicated developmental task, and we can do no more than make the broad outlines of gender-role development here. More than any other aspect of the heterosexual affectional system, gender-role development creates the greatest gaps between humans and other primates, both in the processes involved and in the sex-role contents. Nevertheless, there are some parallels in the sex-role development of all primates, as illustrated by the sex-differentiating behaviors in the age-mate system.

In monkeys the earliest sex differences appear in the patterns of penile erection and thrusting, and threat, rigidity, and passivity. These sexual precursors are early reflections of broader behavioral tendencies which separate male primates from female primates, although bisexual

56

potential is never completely eliminated. Male infants display an increasing preference for high-intensity total-body responses. High-intensity responses lead first to high frequencies of rough-and-tumble play, and as the nipping and wrestling becomes more intense aggression emerges as one of its components. In contrast females display decreasing preferences for rough-and-tumble play and engage increasingly in the moderate-intensity titillating play of chase and be chased. While male monkeys are learning about rank and assertive social roles through aggression episodes, females learn to passively communicate their recognition and acceptance of the brutish potentials of males.

In adults the sex-differentiating patterns of behavior in the peer group lead to the male role of provider and protector and the female role of specialist in nurturance and family, with special focus on child rearing. Monkey social groups are well organized and relatively peaceful, with males selecting the foraging and sleeping locations and protecting the group against destructive intragroup conflicts as well as unwelcome intruders. Unrestrained or frequent aggression, however, does not create tranquility, and a major task of all male primates is learning to manage aggression. The influence of the peer-group affectional bonds, along with the stabilizing influence of social ranks, accomplishes this goal, so that aggression is inhibited among in-group members and is displayed as concerted social action against outside threats.

The role of affectional bonds in socializing aggression is convincingly demonstrated by a study in which male and female monkeys were deprived of affection by raising them in individual wire cages during the ages when infant love and peer love normally develop [Harlow et al., 1966]. As adults these deprived monkeys were tested repeatedly in heterosexual dyads composed of a socially deprived monkey and a normal feral-raised test partner. Threat as a normal heterosexual behavior was used to communicate status and to settle disputes peacefully. Deprived males, however, not only displayed more threatening approaches, but also carried these threats through to damaging physical attacks (Fig. 2-11). These deprived males were totally unable to establish any affectional relationships with normal females because they had developed no aggression control through earlier affectional development.

Females in primate groups are apparently concerned with only the instrumental value of social status as it allows them to select their friends and lovers and provides the other concrete rewards of life. They

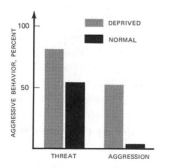

FIGURE 2-11 *Threat and aggression by normal and deprived males.*

typically do not engage in physical confrontations for status. The deprived females in this social-deprivation study did aggress initially against males. Negative feedback, however, quickly suppressed this aggression. Females win their way into male hearts and minds through passive resistance and social sophistication. In our society females usually attempt to combine love and marriage with social security.

In humans these gender-role landmarks are no less important or prominent. Young males prefer action and young females prefer active attention. Males in all cultures studied show greater preoccupation with and display of aggression, and management of aggression could clearly stand some improvement in modern societies. Social psychologists also recognize that social conformity can be more easily created and studied in the laboratory when female subjects are used. Even though child-rearing practices vary from culture to culture, early maternal care in some form is a cultural universal.

Countless studies on sex differences and gender-role development have been conducted on humans. The theories of sex differences and gender-role development, presented with thoroughness by Maccoby [1966], all focus on parental influences in shaping gender roles and claim that gender-role development is the exclusive domain of learning by the child. While learning variables must be heavily emphasized in the ultimate psychosexual outcome, it should be clear that biological behavioral potentials exist as basic primate characteristics. The richness of these potentials merely provides more innate response tendencies which can be shaped into more complex and flexible forms. An interesting theory by Kohlberg [in Maccoby, 1966] concerns the possibility that cognitive maturation provides a biological path to parental influence on children. Parental influences also play a greater role on sex-role development in humans than in monkeys. The advent of a child-oriented society places children under the direct constraint and instruction of adults for most of the daily learning period, and children partially develop the cognitive capabilities for sex-role learning well before their activities are predominantly centered in peer groups. Nevertheless, the peer group does eventually exert powerful forces in shaping behavior unless the capacity for peer love is destroyed. Parental influences on gender-role development should be traced through these age-mate experiences.

The social-motivation component of the romantic subsystem of heterosexual affection consists of the need for sheer and simple social proximity, as this finds a unique expression in heterosexual relationships. This motive for social affiliation has been measured and manipulated in human studies. In humans it appears more readily in females, possibly as a result of pressure toward independence in males. Need for affiliation is always present in greater or lesser degree, although it may be temporarily enhanced by threat or anxiety and temporarily satiated by prolonged propinquity. Affiliation begins at birth in primates in the arms of the mother, and in natural conditions it becomes strengthened through the learned associations in the social group. Deprivation of affectional development in infant monkeys also damages this motive (see Fig. 2-12).

58

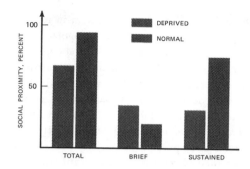

FIGURE 2–12 *Social proximity by normal and deprived females.*

In the monkey isolation study described above, the socially deprived females displayed the usual frequency of brief proximity with normal males, although these proximities were almost totally initiated by their heterosexual partner. However, they avoided any proximity that endured longer than 1 minute. Evidently the failure to experience affection early in life rules out the possibility of later reproductive heterosexual relations.

ABNORMAL SEXUAL PATTERNS

In an absolute sense there are no abnormal heterosexual behaviors. Any heterosexual behavior measured in an adequate sample of any primate, including humans, yields an approximation to a normal distribution. When the heterosexual behavior is sex differentiating we obtain two partially overlapping distributions, one for males and one for females. This overlap indicates bisexual potential, the sexual characteristics present to any and all degrees in the members of both sexes. Bisexual potential is a fact in humans and monkeys, and judgment of any sex-role behavior as normal or abnormal is only in relation to some manmade standard, real or fictitious.

As we saw in Chapter 1, behavioral normality must really be defined in terms of relative frequency of occurrence in any culture. A sexual response is thus normal if it occurs with sufficient frequency in other individuals with the same sexual and social status. Abnormal sexual responses are defined as responses which rarely or never occur in individuals with the same sex classification. On this basis, for example, masturbation is exceptionally normal in male primates and is subnormal, but not abnormal, in female primates. Evidently male primates, by reason of construction and constitution, have a better grasp on this aspect of reality than female primates.

However, statistical normality is not the only standard of heterosexual behavior, and in primates the standard of a male-female affectional bond takes priority. Thus relatively frequent masturbation, in either the bathroom or the bedroom, may be a little bizarre but it represents no intrinsic threat to heterosexual relationships provided that a heterosexual affectional bond has been well established.

59

The greater importance of the romantic subsystem in relation to the mechanical and secretory subsystems is often unrecognized by individuals concerned with the adequacy of their heterosexual adjustments. A prime example is one of the sexual-adjustment problems arising out of the new morality, or the orgasm revolution [Bernard, 1968]. According to recent studies, the current generation of parents has about the same pattern and frequency of premarital, marital, and extramarital sexual responses as yesterday's parents, but they are enjoying it far more than their own parents ever imagined possible—especially the females—as revealed by several measures, including orgasm [Bell, 1966]. Apparently the modern generation has escaped the clutches of guilt associated with orgasmic pleasure. However, cultural fear and guilt are in relation to the prevailing cultural standard, and the present standard is a high level of orgasm. As a result, the previous guilt over experiencing occasional orgasm has been replaced by apprehension over occasional failure to experience orgasm. Thus a perfectly normally functioning heterosexual affectional system can be subverted by application of an arbitrary and isolated standard for secretory sex. Alleged experts on love who equate coital captivity with heterosexual love do not understand the principle of relative frequency, nor have they imagined what life would be without affectional bonds.

We shall not detail here the wide variety of human heterosexual responses which have received value judgments ranging from outright perversion to mere personality problems. The variables operating to produce abnormal heterosexual behavior in humans are almost impossible to sort out from the data presently available since the individuals displaying these behaviors are usually adults when they are first studied, and by this time the bewildering array of variables which have operated simultaneously throughout life are hopelessly confounded. Theories in this area typically ascribe all romantic anomalies to learned variables, and particularly to parental influences on sex roles. Learning, and in part parental influence, clearly does play a significant role. We must assume, however, that learning always acts on some form of biological potential, and that whatever the environmental influence, it operates by interfering with the normal expression and shaping of biological potential in peer-group relations. Let us therefore consider some of the factors involved in abnormal sexual patterns.

Most modern theories of abnormal heterosexual behavior are in agreement that the most severe and irreversible sexual damage occurs when the romantic sex subsystem is destroyed or distorted. The overwhelming importance of this subsystem is borne out by the additional results of partial social deprivation in monkeys. As noted above, in this study male and female monkeys were denied the opportunity to develop infant love and peer love by being reared in individual wire-mesh cages. The heterosexual tests conducted when these monkeys reached reproductive maturity revealed that the secretory subsystem was undamaged by partial social deprivation. Deprived males achieved penile erection and masturbated, sometimes to ejaculation, with normal frequency. Females masturbated with normal low frequency, although orgasm is not presently a measurable response in female monkeys, if it occurs at all. The romantic subsystem, however, was severely affected by partial social deprivation. The need for sheer social proximity was depressed below normal levels, as noted above, and the component of sex-role behavior in the romantic subsystem was also damaged, particularly in males. Although deprived females quickly learned to suppress threat and aggression against feral males, deprived males never developed inhibition of their brutal aggression. The deprived males were sexually aroused by their feral partners, but even when they succeeded in recognizing the female's sex-role postures and solicitations, their only response was puzzlement or aimless groping, sometimes of their own bodies.

The mechanical sex subsystem was damaged but not totally destroyed in deprived females, who displayed the sexual-presentation posture with normal frequency but were unable to support the male adequately when he mounted. Although limited, the partial mechanical support led to the discovery that a major deficit in the romantic subsystem of deprived females was their nonacceptance of heterosexual contact and their basic mistrust of heterosexual contact when it did occur. Deprived females regularly fled when males attempted to mount, and even when mounting was managed on the sly, genital contact was a sufficient threat to convince these females never to leave their flanks exposed. The mechanical subsystem in deprived males was displayed rarely, and only in infantile forms resembling the fragmentary, disoriented presexual postures displayed by young infants during play. Since heterosexual trust and acceptance of heterosexual contact cannot occur at a distance, these components of romantic love could never play a role in the lives of the deprived males.

Figure 2-13 summarizes the coital consequences of partial social isolation in male and female monkeys. These results show that while social deprivation produced subnormal frequencies of mechanical sex in both males and females, the effects were far more severe in the deprived male. This pattern of greater heterosexual crippling in the male than in the female was found in virtually all heterosexual responses measured.

The greater effect of experiential factors in males than in females has been confirmed in a variety of test situations with many different species, ranging from rodents to carnivores to primates [Beach, 1947]. The human male has also been described as more susceptible than the

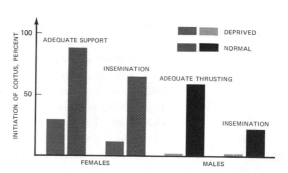

FIGURE 2–13   *Sex differences and partial social deprivation.*

human female to sexual developmental experiences, although the data used to support this sex difference in humans is based on biased samples and to some extent it undoubtedly measures differences in bias. It would not be surprising, however, to discover that human males are more sensitive to the operation of early social experiences than human females. As Money [1970] points out, the basic theme of nature is to produce a female, beginning prenatally and extending throughout the period of reproductive immaturity. Prenatally, a fetus develops with bisexual potential until the gestation period for sexual differentiation. At this time a female fetus develops unless specific hormones appear to create embryological differentiation into a male. Specific hormones must be added—and in "genetic" males they almost invariably are—to create a male fetus. The presence or absence of these hormones determines whether the individual will have the anatomical and physiological characteristics of a male or female. In monkeys the presence or absence of these hormones also influences whether the early sex-differentiating behaviors, such as threat responses and play-initiation responses, will be masculine or feminine in form. In humans the influence of prenatal hormones on postnatal differences in sex behavior is certainly less prominent, if it occurs at all.

A second developmental shift for males occurs at the stage when the male child must loosen the infant-mother bond and begin the development of male-appropriate behaviors of assertiveness and independence. The female child at this stage can maintain the infant-mother bond to a greater extent without cultural criticism or threat to her subsequent femininity. A third developmental difference between males and females may occur at adolescence, when the male's total devotion for intense athletic activities with boys alone must be tempered by tender loving manipulations with the allegedly frailer sex. Adolescent girls, in contrast, probably experience few problems in transferring their play with other girls to the nurturance and management of heterosexual affection.

Whether these basic development differences between males and females will shed any light on sex differences in heterosexual difficulties is a topic which must await the revisions of this text. However, other sex differences have been uncovered in the adult which may contribute to our understanding of the different heterosexual adjustments required by males and females. Males are apparently more capable of sexual arousal through the distance receptors than are females. Males, more than females, report sexual arousal when they are presented pictures and narrative material with sexual content. If this effect proves to be a reliable and biologically based sex difference, it will have implications for heterosexual behavior ranging from abnormal sexual fixations to panty raids. The most basic sex difference, the difference in anatomical location of the genitalia, may also create different adjustment problems for males and females. Bernard [1968] reminds us that the incompletely aroused male has no choice about disclosure of detumescence, whereas the female can and often does fake arousal and orgasm, usually to protect the male against any feelings of inadequacy he might experience about his inability to bring her sexual fulfillment. The incompletely aroused male must hide his sexual performance in shame, while the incompletely aroused female can hide her sexual performance in love.

The paternal affectional system, tenderly illustrated in Fig. 2-14, is the affectional relationship of an adult male for an infant. The infant may be either a male or a female, and the infant may reciprocate with a greater or lesser degree of infant-adult male affection.

PATERNAL LOVE

Biological kinship between the adult male and the infant is not a defining characteristic of paternal love, since substitute fathers are not infrequent—and may even be a great deal more frequent than they think.

The paternal affectional system and the variables which affect it have been far less adequately studied in both monkeys and in humans than the other affectional systems we have discussed. While the data are limited, they nonetheless suggest that innate biological variables are minimal in paternal affection, and experiential variables are maximal, along with a great degree of cultural determination of human paternal love. The relative absence of innate biological paternal potential would imply, of course, that the paternal affectional system was not designed to serve an essential biological function. In nonhuman primates, when it occurs, it does often serve as a secondary protective system for infants. Paternal love occurs so irregularly in monkeys and apes, however, that this function appears to be little more than a coincidental consequence of learning to love and protect a baby through frequent contact with the baby's mother. Mitchell [1969] has summarized the studies of paternal affection in nonhuman primates, and the variables suggested by field studies strengthen the view that factors which create painless propinquity between an adult male and an infant create favorable conditions for learning a protective paternal passion. Adams [1960] suggests that the nonbiological paternal affectional system results from the conditions created when the two essential biological dyads, the maternal dyad and the heterosexual dyad, function together. This model was designed for the

FIGURE 2–14   *Paternal love.*

human paternal system, but it also works well in organizing the variables which influence the paternal affectional system in monkeys and apes.

The paternal affectional system in monkeys has never been found with the length or strength in which it occurs in some cultures, such as our own. Adams thus maintains that the paternal dyad in humans, as in monkeys, is neither required nor designed to serve biological functions. In some primitive societies the maternal dyad and sexual dyad operate apart from the nuclear family arrangement of father-mother-children to provide the characteristic societal functions. In other societies, where there are different patterns of economic, educational, and socialization concerns, the nuclear family arrangement, or some other, may be an essential factor in carrying out these concerns. However, analysis of the functions and the variables underlying paternal love in primates probably should not be oriented toward a search for the biological variables and the biological roles of paternal love. Perhaps the paternal affectional system is unique in that it is learned along different dimensions to serve different functions in different societies.

Love, more than any other emotion, is characterized by the properties of both the devil and the divine. Since love encompasses such diverse facts and faces, it carries with it unlimited possibilities for frustration as well as fruition. Torn between love's potentials and penalties, we harbor thoughts and feelings about it which we often find difficult or impossible to express, let alone examine objectively. Although we look upon our own love as a gift, the framework of the laboratory setting enables us to view the multitudinous aspects of primate love in general with a greater measure of objectivity.

SUMMARY

It is primarily through the age-mate affectional system, expressed in peer play, that social and cultural patterns are learned, control of aggression is accomplished, and the foundations are layed for later sex-appropriate behavior. The stage of presocial play, characterized by exploration, parallel play, and instigative play, provides progressive preparation for interaction with peers. The stage of social play takes three forms—free play, creative play, and formal play. Free play, clearly the most important socializing influence, is typified by rough-and-tumble play and various forms of approach-withdrawal play which provide learning in social ordering, development of later sexual posturing, and the formation of specific affectional relationships. Creative play, typified by the use of objects for purposes other than their original or primary ones, extends to the creative efforts of adulthood. Formal play, characterized by the fixed rules of formalized recreational activities, also extends to adulthood.

The heterosexual affectional system develops in all primates through three discrete subsystems. The subsystem of mechanical sex begins in infancy with reflexes of penile erection and pelvic thrusting and develops through the gender differentiation and presexual posturing of the age-mate period into the complete behavioral pattern of adult coitus. The secretory subsystem develops with the maturation of hormonal influences on secondary sex characteristics and behavior. The romantic subsystem, which is based on an affectional bond between two members of the heterosexual dyad, is an outgrowth of the preceding affectional systems by transitional mechanisms that lead to the four basic requirements for normal heterosexual functioning—heterosexual trust, acceptance of contact, behavioral gender differentiation, and motivation for social proximity. The romantic subsystem occurs in all primates and in some other animals, but in man additional transitional mechanisms are probably required to provide for the enormous cultural, cognitive, and individual complexities engendered by man's capacity for thought and language. Abnormal sexual patterns are generally a result of improper development of the romantic subsystem. Although secretory sex is not affected, the failure or destruction of antecedent affectional systems, especially the peer system, damages not only the romantic subsystem, but the mechanical subsystem as well, with more severe effects in males than in females.

The paternal affectional system often serves as a secondary protective system for infants, but it varies widely with species and culture, and its roots seem to lie in experiential rather than biological variables.

CHAPTER THREE

AVOIDANCE AND AGGRESSION

Human beings are born with the face of the devil as well as the face of the divine. We have already discussed love in its five aspects. The twofold face of the devil appears with the advent and development of the evil emotions, fear and anger. In a strict operational sense fear and anger are emotions, not behaviors. However, associated with these emotions are sets of behaviors which can be defined and delineated by observation. A major consequence of the emotion of fear is the expression of fear-elicited behavior. The emotion of anger has a similarly clear behavioral expression. In general fear is expressed in avoidance of the fear-eliciting person, object, or situation, while anger takes the form of approach behavior directed toward the infliction of psychological or bodily harm—aggression.

It is paradoxical that social integration is determined in large part by the antisocial motives of fear and anger. However, there are two reasons for this phenomenon. First, the emotions of love are ordinarily prepotent over anxiety, anger, and even pain. Therefore the positive, integrating forces of love can operate effectively to shape the individual and his society, regardless of the behavioral consequences of the evil

FEAR AND ANGER

emotions. Second, love matures before the evil emotions mature and operates to ameliorate or channel the evil emotions. Unrestrained aggression might well endanger the survival of any species with well-developed social structures, but controlled or socialized aggression in these species is essential to the development of cohesive groups which work together against outside enemies. Thus aggression operating through the channels provided by antecedent love increases the probability of survival for the species.

There can be little question that with respect to species which exhibit complex social behavior, particularly primates, the potential for fear-elicited and aggressive behavior is inherited. It is inevitable that any social organism will exhibit these responses throughout its existence, regardless of the environment in which it is raised. However, the form and frequency of such behaviors are shaped by environmental influences, social and nonsocial. Both the certainty of occurrence and the flexibility of expression are readily observable in maturation studies of human and nonhuman primates. The natural history of aggression and the neural mechanisms that produce it are considered further in the context of motivation.

Data from a series of long-term experiments with rhesus monkeys

67

lead us to believe that the nature and operation of social learning depend in large part on the orderly maturation of the three basic social-emotional patterns of affection, fear, and aggression. Some form of affection—positive adient response toward members of the species—precedes widespread fear responses toward specific external stimuli of less than catastrophic physical intensity, and specific fears in turn precede severe and physically damaging intraspecies' aggressive responses. In the rhesus monkey these patterns mature in orderly sequence. Affection is normally well established by two months, signs of fear emerge at approximately three months, and signs of aggression are nonexistent before the age of six months. Furthermore, the normal progression of this sequence is essential to all later stages of development. Any environmental manipulation that distorts, disturbs, or otherwise denies this orderly sequence leads to maladjustment in later social-sexual development. In other words, the antisocial responses of fear and aggression are socially constructive for the individual, and ultimately for the species, only when they are preceded by the normal maturation and development of love.

FEAR AND ANXIETY

Like all basic emotions, fear arises from both unlearned and learned variables. Even though we know that all such emotions are primordially unlearned, it is sometimes difficult to show when they first arise in newborns of any species and also to define the stimuli that elicit them. The newborn human has little voluntary control over his musculature, and the few gross responses he can make are doubtless inadequate to express his full emotional repertoire. After extensive observation of many infants Bridges [1932] proposed that there were definite stages in the differentiation of emotions. She ascribed to the newborn only excitement, which differentiated at three months into delight and distress. Even before six months of age, distress differentiated into disgust, fear, and anger, and a short time later delight differentiated into elation and affection—affection first for adults and later for other children. There is reason to believe that the newborn child differentiates and responds to pain and pleasure long before three months and probably differentiates fear and anger before six months, but the pattern of differentiation is essentially the same for all infants.

Although the role of learning in the acquisition of specific fears and targets for anger is obvious, this evidence of a specific sequence of emerging emotional states enables us to reinterpret certain early theories concerning the hereditary basis of emotions. An infant does not, as was once believed, inherit the specific behavioral expressions of emotions that were exhibited by his ancestors. Rather, the capacity for certain emotional states emerges through maturation, and through learning the number and nature of stimuli that elicit reactions in the individual expand. Similarly, emotional behaviors are acquired not during embryonic life, but through learning based on antecedent maturation processes. A joyful mother may well produce a joyful child. However, this is more likely to be a result of imitation as the growing child constructs his picture of reality than the mother's joyful state during pregnancy. The mother's emotional state does, of course, influence the fetus. Emotional states in the mother are accompanied by changes in the autonomic nervous system, which affects hormonal secretions from various glands and influences the chemical composition of her bloodstream; these chemical factors are transmitted through the placenta to the fetus' circulatory system. Thus, although there is no direct connection between the nervous systems of mother and fetus, the mother's emotional state during pregnancy may clearly affect the general overall development of the fetus. Specific emotional reactions, however, are not transmitted from mother to fetus.

The first psychologist to investigate the fundamental stimuli that elicit emotional responses was Watson, who postulated that the basic emotions of fear, anger, and love were innate, and that each could be elicited separately at birth by a specific stimulus category [Watson and Rayner, 1920]. He contended that fear was elicited by loud sounds and loss of support (falling), anger by physical restraint, and love by stimulation of the erogenous zones—a kiss full blown to Freud. Although the foundations for these three emotions may well be innate, they are not discernible independently at birth. Watson's classification of eliciting stimuli has proved to be basically accurate. However, his theory does not include the possibility that the emotional differentiation that develops at later stages may be a result of maturation as well as learning.

ANXIETY-ELICITED BEHAVIORS

Two very prominent examples of emotional-response patterns which emerge through maturation at different ages are stranger anxiety and separation anxiety. Some time during the infant's sixth month a stranger's face elicits a reaction termed *stranger anxiety*, which is evidenced by tightening of the infant's face and the initiation of crying. The reaction ceases as soon as the stranger moves out of the infant's line of vision and is elicited again when he reappears. Fortunately for harassed parents, by the twelfth to fifteenth month this reaction has usually disappeared [Schaffer and Emerson, 1964a]. However, what is it in this situation that elicits anxiety? Anxiety is the typical emotional response to a situation which is perceived as novel or unfamiliar. By six months the child has sufficiently developed a schema or mental image of his mother's face that he is readily able to detect discrepancies from it. Hence at this age a feeling of anxiety follows awareness of a new or unfamiliar face. Undoubtedly the infant's schemata later expand to permit a greater tolerance for unfamiliar faces. Eventually the unfamiliar even offers delight and intrigue.

Another situation which often elicits anxiety in the six-month-old is a feeling of helplessness. When the normal flow of events as perceived by the infant is disrupted there is little he can do to alter the situation and his only available response is anxiety. This emotional reaction has been summarized by Mandler [1964], who has emphasized the fact that the inability to complete a sequence of activities and the lack of alternate completion sequences results in helplessness. Furthermore, he believes that helplessness is, by definition, anxiety.

In specific reference to stranger anxiety, the six-month-old is clearly limited in his ability to manipulate and control his environment. Stranger anxiety may subside during the following months and be totally absent by the age of two. At this age a stranger may indeed pose potential threat, but the two-year-old is more capable of alternative responses in order to control circumstances. Not only can he run to his mother, but he can also ask questions about the situation, which may include information about the stranger.

Another anxiety, termed separation anxiety, emerges at approximately ten months. This is evidenced by an increased tendency for the infant to cry whenever his mother leaves him. This protest reaction to temporary separation from the mother is not exclusive to human infants. Monkey infants exhibit similar reactions, characterized by marked increases in infant vocalization and agitation. As we saw in Chapter 1, an infant monkey placed alone in a strange situation will typically clutch himself, huddle, and scream with fright. However, in the safe presence of his mother, real or surrogate, the same situation fails to elicit any fear at all. In time the infant will actually search out and explore the novel environment, using the mother as a base for security. The object of separation need not be a mother. When a six-month-old monkey raised with only a cloth diaper as a companion is separated from his diaper, the same protest reaction is elicited. Protest reactions to separation and separation anxiety thus appear to be innate mechanisms which develop after the infant has formed an attachment to an object—any object—to which he has been exposed for a substantial period of time [Cairns, 1966].

THE LEARNING OF SPECIFIC FEARS

Doubtless most of our adult fears are learned, but the type of learning that produces fears also changes with age. Early in life most fears are probably learned by a conditioning process—the association of a new stimulus with an unlearned stimulus and generalization of the new stimulus to objects physically similar. The acquisition of fears by conditioning was first demonstrated by Watson [1920]. He began by showing a young infant, Albert, a toy white rat, which initially evoked no fear responses. Next, every time he presented the rat he also presented a loud sound, a stimulus he knew would evoke a fear response in young infants. It took only six or seven trials for Albert to exhibit fear toward the toy rat as well as the loud sound. Furthermore, Albert's fears gradually generalized to similar objects such as a cotton ball and a white beard, and the more similar the object to the rat, the more fear it evoked. Thus fears may be learned very early in life, certainly during the first year, and may persist long after memory of the event that caused the fear is lost. In addition, as indicated by Albert's behavior, fears tend to generalize very broadly, especially when they are formed early in life.

As soon as the child forms social attachments to his mother and father, and subsequently to playmates, he learns fears through social imitation, and there is a relatively high correlation between the nature and number of fears of mother and child. For example, during thunderstorms the tense and perturbed parent may unwittingly serve as a model for the growing child, who imitates the behavior of those closest to him. At later ages children may acquire a large number of fears as a result of language techniques and by sheer fantasy and imagination. Once formed, fears are very persistent. Many fears formed at the age of five persist past fifty. Even adults who are generally described as brave frequently have one or more persistent, disparate fears or phobias. Furthermore, fears are often formed without the desire or effort to learn them and are frequently acquired against the conscious or unconscious wishes of the victim.

As children mature their emotions become more specific and often deeper in their expression. Moreover, the stimuli that arouse specific fears are different at different stages of development. Before the age of two noises may account for as much as 25 percent of the fear-eliciting stimuli, whereas by age twelve only 3 percent of the child's fears can be attributed solely to agents of noise. There is a similar decline with age in fear of strange situations and strange persons. Imaginary creatures play an increasing role in the arousal of fear as the child grows older. For the young child, ghosts and goblins lie lurking in the darkness. An astonishing 20 percent of the fears held by young children deal with such imaginary creatures [Jersild et al., 1933].

Many childhood fears are just outgrown, especially if the environment offers a feeling of security and nonadaptive fears are actively discouraged. Intense fears, however, do not dissipate with time, and the customary techniques employed to eliminate them are by no means equally effective. Jersild and Holmes [1935] found that ridiculing or ignoring a child's fear was essentially ineffective in eliminating it. More drastic action such as physical punishment was found to be equally ineffective. Demonstration, explanation, and reassurance are substantially better, but not entirely successful. However, if a pleasurable association with the feared object can be established, the original fear may be overcome by *counterconditioning*. Watson found that Albert's fear of the toy white rat was gradually removed by presenting something pleasurable, such as candy, with the toy. An outgrowth of this technique is the behavior-therapy method of desensitization.

71

Many fears, of course, are healthy and adaptive in that they serve to protect the individual. Parents have every hope of instilling realistic fears of ocean undertows, busy intersections, and loaded guns. However, fears instilled to the point of obsession may limit the horizons of the growing child, who for maximal personal development should be in the process of exploring his environment. An obsessive fear toward one situation may gradually be generalized to other situations in which no fear is warranted and may even lead to a general state of apprehension or anxiety under all circumstances. In extreme cases anxiety can be totally disabling and result in the eventual elimination of all environmental and social interaction.

Perhaps the most frequent consequence of nonadaptive fear reactions is their generalization to a stage of *free-floating anxiety*. Here the emotional reaction is not in response to specific objects or situations. Instead the individual is burdened with feelings of apprehensiveness, uneasiness, and finds himself full of foreboding fears. Although there are exceptions, the emergence of anxiety in the child can often be traced to either direct or indirect parental factors and forces. In addition to the simple imitation of apprehensions in one or both parents, parental attitudes and child-rearing practices themselves may initiate feelings of anxiety. After extensive research Saranson [1960] has suggested that one antecedent condition leading to anxiety is the particular parent-child interaction in which there is a constant parental threat of negative evaluation of the child's performance. The one ready reaction to this negative evaluation is aggression. However, since the child is dependent on adults, the conflict that arises between his dependency needs and this aggression reaction may lead to anxiety. Saranson also reported that mothers of highly anxious children tended to evaluate their child's behavior in terms of unrealistic standards of his capabilities. As anxiety affects the adult, so it affects the child.

Anxiety is an uncomfortable experience, and the individual often puts forth considerable energy to avoid or alleviate this distressful state. In the sense that defense mechanisms provide protection against painful experience, they serve an adaptive function. However, in cases of more extreme anxiety the rigid and inappropriate behavior adopted as a defense can be self-defeating in that it becomes a source of further anxiety and eventually interferes with adaptive function. Thus in situations where initiative, creativity, or utilization of complex cognitive processes are required, anxiety takes its toll by inhibiting effective performance and fruitful pursuits.

Since fear and anger both are socially disrupting emotions, all primate social groups have evolved adaptive mechanisms of emotional control. Fears, particularly early fears, are held in check primarily by the antecedent formation of infant love, the formation of basic security and trust by the infant for the mother. In the mother's absence the infant monkey fears everything; it is a baby in terrible terror. Although the primary security safeguard is the mother, this function is extended by age mates and adults with chronologic development.

72

In the human infant anger becomes a distinct emotion when, along with fear and disgust, it is differentiated from the more general emotional state of distress. The human infant is allotted less than six months in his new world before he begins to experience this emotion many times, for many reasons, and he will express it in many ways all his life. The initial manifestations of anger, however, become modified very early in life. Goodenough [1931] noted that expression of anger changed from temper tantrums and outbursts of uncontrolled motor activity of the infant less than one year of age, to the directed motor and language responses of the child by the age of two. These latter responses had accounted for a mere 14 percent of the outbursts of the younger infant.

Although the antecedent conditions of anger in the adult usually involve psychological or social limitations or frustrations, the conditions leading to the arousal of anger in the very young are primarily physical. Restrictive clothing, confinement, and denial of desirable play activities are all situations which can give rise to anger in the child. Obviously some restrictions are necessary for the survival of the child and the sanity of the parent. However, many such restrictions are all too often based primarily on arbitrary rules or ungrounded parental fears. Arbitrary and inconsistent parental demands may present the child with an unsolvable, and hence frustrating, discrimination problem. Such feelings of frustration may be vented through aggressive behavior. Aggression, however, may be directed either inward, toward the self, or outward, either toward the members of one's own group or toward members of other groups.

Self-aggression can be observed in rhesus monkeys raised in social isolation. This behavior becomes prominent after three years of age and is frequently exhibited when strangers are present in the colony room. These angry monkeys chew on their own hands, arms, feet, or legs, sometimes to the point of tearing the flesh (Fig. 3-1). It is somewhat similar, in the anthropomorphic sense, to the self-destructive behavior displayed by some autistic children. Self-aggression in humans is often reflected in psychological disturbances, with the most extreme form being suicide. This violent form of self-aggression is a complicated aspect

FIGURE 3–1   *Self-aggression.*

of human life, and no simple variable can account for its various manifestations. For example, suicide rates are not constant across the United States. Urban-dwelling persons commit suicide more frequently than do rural-dwelling persons, with the greatest incidence in the most densely populated areas. Single, widowed, or divorced people have more suicidal tendencies than married people, and although more females attempt to take their lives, males have a higher rate of success. Also, the suicide rate increases with age.

Innate aggressive tendencies emerge through maturation at different ages, depending on species, sex, and individual differences. There is little disagreement among comparative psychologists that aggression is part of the biological heritage of primates. However, some social psychologists who limit their studies to the human animal still believe that aggression is basically a learned behavior, and that the differences which occur between the sexes or among individuals within their sex group are accountable solely on the basis of experience. No doubt the late appearance of aggression in the developmental sequence has led some observers to underestimate its biological basis. It is customary to accept maturation as a factor in the development of locomotor behavior, language, intelligence, and sex. However, with respect to human social behavior, which has a large and obvious learning component, there is apparently reluctance even to conduct the research that might demonstrate a maturational variable, and studies have focused on the variables that alter the expression of aggression after the behavior has already developed.

As we saw in Chapter 2, aggression not only has clear precursors in both monkey and human play long before it matures as a behavioral expression, but in the monkey there are innate gender differences which play a vital role in later heterosexual development and behavior. Although the specific patterns of behavior are different in the human being, these monkey data lead to the speculation that the sex differences observed in human aggressive expression result not from learning alone, but from learning superimposed on similar innate differences in aggressive tendencies. Learning undoubtedly exerts considerable influence in our culture, where young males are often actively encouraged by both maternal and psychological rewards to behave aggressively and young females are usually reprimanded for action which is aggressive. Given these conditions of reward and punishment, it is not surprising that the two sexes quickly develop exaggerated differences in the display of aggressive behavior. Increasing age produces not only greater sex differentiation, but also qualitative changes in aggressive behavior. Screaming, weeping, and physical attacks, evident in the very young child, decline with age as verbal aggression increases.

The emerging sex differences are also reflected in terms of the stability and continuity of aggressive behavior. Data from a longitudinal study indicate that aggression is much more stable in males than females [Kagan and Moss, 1962]. It was found that the rage and temper tantrums displayed by the male preschooler were more predictive of his later aggressive behavior and the ease with which he expressed anger in adulthood than was similar behavior in the female preschooler. Presumably the lack of punishment for aggression encourages the perpetuation of such behavior in the male, whereas aggressive young females soon learn through social pressures to inhibit overt expressions of aggression.

THE SOCIALIZATION OF AGGRESSION

Because the emotion of anger develops in all infants, and one outlet for this emotion is aggression, the socialization of aggression is a primary concern for any social group. Socialization of aggression does not mean the total inhibition of aggression, even in females. The aggressiveness of any male could not surpass maternal aggression when the life or safety of her young is threatened. Such behavior is adaptive for the survival of both the individual and the species. However, harmful aggression between members of a species must be controlled if the species is to survive. The techniques of control vary, of course, with the species.

In the rhesus monkey the maturation of positive affective feelings precedes aggression and operates to channel aggressive behavior. Thus when aggressive behavior does appear, usually at approximately eight months, it is ameliorated primarily through age-mate play behaviors. In the second year aggression becomes a commonplace social behavior for males. Not only is there direct aggression but displaced aggression as well, with clear-cut scapegoating. A common observation, both in laboratory and field situations, is that an animal intermediate in the dominance hierarchy of a social group, when attacked by a more dominant group member, will subsequently attack a less dominant member of the group with no apparent provocation; this is also a commonplace human phenomenon.

Part of socialization entails the learning of appropriate targets for aggressive tendencies, and the monkey must be reared in a group during the period when aggression matures in order to learn such targets. Monkeys reared only with their mothers for the first eight months of life are hyperaggressive when subsequently exposed to peers. Monkeys reared in total social isolation subsequently exhibit aggressive behavior, but it is exceedingly ill-directed. Such an animal may attack an infant, something a socially sophisticated monkey would never do, or attempt to attack a dominant male, something few socially reared animals are stupid enough to try. Whereas social isolates will often aggress against themselves, self-aggression is exceptional behavior for a socially sophisticated monkey, even under unusually stressful environmental conditions. Although aggressive behavior is inevitable for maturing and adult monkeys, it is obvious that its occurrence and direction are derived from social experience.

By the time anger has developed into a powerful social or anti-

75

FIGURE 3–2 *Monkeys collaborating against a crocodile.*

social force the mother-infant bonds have long since weakened. For this reason age mates become the primary social source of aggression control. Through peer play a new affectional bond develops which tends to hold aggression at a controlled, nonlethal level. The age mates with whom we play, in fact or in imagination, become our in-group, and all others serve as out-groups. Aggression against out-group members is strong and can be augmented by a host of environmental variables to the point of lethality.

Aggression control and cooperation by in-group members of a monkey clan are illustrated in Fig. 3-2. The monkeys on the island are a cohesive in-group, and those who play together slay together—not each other, but the out-group members, who in this case are innocent crocodiles, given that crocodiles are ever innocent. Working in happy cooperation, the monkeys are preparing to pull the crocodile against the cement wall of the moat in order to chew on its soft underbelly.

Although aggression control is achieved primarily through age-mate affection, there is no question that parental training exerts powerful influences, usually for the better and occasionally for the worse. In human socialization, of course, the young often learn to control their parents' behavior instead of their own. The child whose parent succumbs to temper tantrums quickly learns that aggressive behavior can be an effective operating procedure and continues to employ it as a means of gaining his own ends.

MODELING OF AGGRESSIVE BEHAVIOR

Human parents exert considerable influence on the child's containment and acquisition of aggressive behavior. This influence is often exerted as a conscious attempt to control undesired behavior, but probably just as often they are unaware of their influence as available models for imitation. There is evidence that spanking, slapping, harsh verbal reprimand, and other methods of punishment lead to an inhibition of overt aggression in the home and places similar to the home. Studies by Sears et al. [1953] of aggressive behavior in nursery school children indicate that the overall effect of parental punishment is actually fairly complex. Those children who had mildly punitive mothers exhibited a relatively high number of aggressive responses, while those children who had either severely punitive mothers or nonpunitive mothers showed relatively few aggressive responses.

Although both the latter groups of children did not exhibit overt aggression, Sears argued that only the children of nonpunitive mothers were relatively free of aggressive feelings. He suggested that if aggressive reactions were inhibited by parental restrictions, the child would be more likely to express them in a fantasy situation that differed from the structured nursery school setting, which in many ways resembled or was associated with his home environment. In a permissive doll-play session the children of severely punitive mothers exhibited a relatively high number of aggressive responses. The investigators concluded that these children were actually highly frustrated and experienced aggressive feelings which they inhibited both at home and in other places similar to the home.

Although parental punishment may inhibit aggressive behavior at home, it also plays a much more complex role. Parental punishment, particularly physical punishment, provides the child with an excellent demonstration of how powerful an influence aggression can be (Fig. 3-3). Since it is effective in his case, he may very well apply it to others. Although the eventual results may be unfortunate, the discrimination problem is not a difficult one: inhibition of aggressive behavior at home avoids punishment and aggressive behavior away from home brings success—at least for a while.

The profound influence of an aggressive model on subsequent behavior has been demonstrated in a variety of experimental settings [Bandura and Walters, 1963]. Aggressive tendencies in children were observed after exposure to aggressive and nonaggressive models. The aggressive model engaged in highly novel aggression to ensure that the

FIGURE 3-3 *Monkey infant at the hands of a punitive mother.*

FIGURE 3–4   *Imitative aggressive behavior*

modeled behavior could be clearly differentiated from previously exhibited aggressive behaviors (see Fig. 3-4). The findings overwhelmingly indicated that children who watched aggressive behavior in a model engaged not only in more imitative aggressive behavior, but also in more total aggressive behavior. In other words, an aggressive model may serve both as a transmitter of new patterns of behavior and as a stimulus to elicit previously learned aggressive behavior. Furthermore, those children who viewed a nonaggressive model performed fewer aggressive acts than did those in a control group who viewed neither model. Apparently an appropriate model, parental or otherwise, can also serve as an inhibiting agent for aggression.

Live models were used in the original study. However, in a subsequent study live models, filmed human models, and cartoon models were used to test for differential effects. Although the live model elicited more subsequent imitative aggression, there was no appreciable difference in the total aggression elicited by models of different forms. This finding has some bearing on the current controversy over the possible effects of violence on television. The question of whether violence begets violence, or whether viewing violence in others has a purgative effect on aggressive tendencies has long provided a battleground for proponents of Freudian theory and those who attach major importance to the influence of modeling. According to the Freudian position of *catharsis*, the individual possesses a limited amount of "aggressive energy," and if he views aggressive behavior in a model, he will "use up" some of this limited supply and hence will be less likely to engage in aggressive behavior than one whose supply of aggressive energy is still untapped. Many social learning theorists argue that the converse is true—observation of aggression is likely to increase rather than decrease the probability of aggressive behavior. The practical questions arising from these theoretical differences of opinion are of great import to our media-oriented society. Do television and movie scenes of destruction provide us with a "safe" way of venting our own destructive tendencies? Bandura's model study clearly shows that film models are extremely capable of eliciting substantial aggressive behavior from children viewing the film.

Additional data on college students strongly support the contention that viewing aggression begets aggression in the observer [Berkowitz, 1968]. It has been found that merely viewing objects such as guns associated with violent acts can serve as effective stimuli for triggering impulsive aggressive acts. In one experiment half the subjects were first subjected to a condition of humiliation and physical discomfort by a confederate of the experimenter posing as a naïve subject. This condition was designed to create feelings of anger in the subject. The remainder of the students, who were not subjected to this condition, constituted the nonangry group. Subjects in both groups were then given an opportunity to express aggressive feelings. Each subject was to indicate any rejection of an idea suggested by his partner by administering an electric shock. Of course the "partner ideas" were prearranged and were the same for all subjects. Half the students in each group were exposed to weapons, guns, and rifles, which simply lay in the experimental room; tennis racquets were present for the other students. The mere presence of guns apparently triggered the release of aggression for those persons in the angry group. These subjects administered significantly greater numbers of electric shocks to their partners than did angry subjects exposed to the tennis racquets.

To test whether aggressive tendencies may be vented through a safe avenue simply by viewing another's aggressive acts, the experiment was repeated, but with a slight change. This time half the subjects were shown a violent film and the other half were shown a nonviolent film, after which all were subjected to a humiliating and uncomfortable condition. The results generally showed that subjects exposed to the violent film administered a greater number of electric shocks to their partners than did subjects exposed to the nonviolent film. Clearly the subject's feelings were not purged through viewing aggressive acts; on the contrary, such scenes seemed to justify the release of his own aggressive feelings.

Numerous other experiments have similarly failed to disclose any evidence of a cathartic effect. Rather, the overwhelming majority of these studies have demonstrated that in fact the opposite effect is likely. The catharsis hypothesis has compelling intuitive plausibility, but it serves as a frightening example of discrepancy between social ideals and social reality.

DISPLACEMENT OF AGGRESSION

Sometimes the target for aggression is not known or is repressed, or a direct attack would engender undesirable consequences. In such cases aggression may be *displaced*, often toward those not capable of effective retaliation. Once a "safe" target for aggression is found, it often becomes a scapegoat for any and all feelings of frustration. Displacement of aggression, or scapegoating, may be on an individual level—as when a man berated by his employer goes home and berates his wife—or it may take place on a social scale—as when one cultural or ethnic subgroup becomes the target for a large segment of society. However, displacement of aggression is not exclusively a human trait. Monkeys and mice displace displeasure as readily as men. Although displaced aggression in monkeys

79

is often viewed with tongue in cheek by man, human prejudice and scapegoating are seldom so humorous.

In spite of the fact that social-emotional control is determined in large part by learning, and in man by culture, the efficacy of this learning is made possible by maturational forces. Infant-mother love develops before external fears become overwhelming, and age-mate affection is firm and fast before aggression has become a powerfully pervasive force. There is a wealth of data supporting the position that both fear and anger have fundamental innate components and develop over a considerable period of time, probably throughout puberty. Although the potential for fear and anger is inherited, most specific fears and angers are learned—in man, monkey, or mongrel. In man emotional learning is so pervasive that the existence of basic unlearned variables may be obscured, even such important variables as those associated with gender differences. It is in the laboratory that the true role of these unlearned components has been unmasked.

SUMMARY

The behavioral expression of the emotion of fear is generally avoidance of the feared object. Anger is usually evidenced by some form of aggression. The potential for fear is innate in any species that exhibits complex social behavior, but specific fears and specific targets for aggression are determined by learning. Despite the fact that fear and anger are fundamentally antisocial emotions, they are channeled by the prior maturation of affectional feelings, and play an important role in social integration.

Fear and anger follow a basic developmental sequence. The first emotion displayed by the infant seems to be a general excitement, which differentiates at about three months into delight and distress; by six months distress has differentiated further into disgust, fear, and anger. A short time afterward delight differentiates into elation and affection, which is directed first toward adults and later other children. There are reasons to believe that the newborn child also responds differentially to pain and pleasure before three months and can probably differentiate fear and anger well before six months. Emotional reactions in the newborn are elicited almost as reflexes by a limited number of stimuli. With maturation other stimuli become signals for emotional reactions, as in the development of stranger anxiety and the later appearance of separation anxiety. Anxiety is also provoked in the infant by feelings of helplessness.

Children acquire early fears largely by conditioning. Later social imitation, imagination, and other more complex learning processes contribute to the formation of specific fears. Many childhood fears are outgrown, but intense fears do not always dissipate with time, and they may be generalized into free-floating anxiety. Sometimes intense nonadaptive fears or phobias can be overcome by counterconditioning. Although fear is a socially disrupting emotion, it is held in check by the basic security and trust formed during the infant-love period and extended during the age-mate period.

Anger is elicited in the infant by physical restraint and later by various social limitations. Where there is no outlet for the resulting aggression, it may be turned inward, as in the self-aggression that characterizes autistic behavior and suicide. Aggression that is turned outward may be directed against either the individual's own group or against other groups. Although aggression within the social group usually has unfortunate consequences, to the extent that aggression against out-groups serves to protect the in-group it has an adaptive function.

All social groups have evolved mechanisms for the control of aggression. The mother and infant affectional systems serve initially to curtail aggressive tendencies. However, peer play is the primary means of aggression control, for it is in the interaction with peers that the appropriate targets for aggression are learned and misplaced aggressive responses are punished. Aggressive patterns are also learned from parental and other models. In some cases aggression is displaced, generally to those less capable of retaliation.

CHAPTER FOUR

MAN: A
SOCIAL ANIMAL

In the most general sense the study of man is the study of how we human beings relate, or fail to relate, to each other. Hermits are clearly in the minority in our society. Most of us, by necessity and by choice, have daily dealings with other human beings. In our daily activities we talk to and listen to others, we think about others, make plans to be with others or sometimes to avoid them. We love, we hate, and sometimes we kill other human beings. Even when we are alone our behavior is constantly guided by social factors. We eat socially acceptable foods, wear stylish clothes, and use socially proper drugs. What we mean by "acceptable," "stylish," and "socially proper," of course, varies. What is "groovy" or "cool" for a teenager is sometimes regarded as inappropriate by his parents—a fact that also guides much behavior. Thus, in one way or another, most of our behavior is influenced by what we think others will think about the way we behave. Our self-esteem depends on our appraisal of the reactions of others to us. Our feelings of security and accomplishment or of insecurity, inadequacy, and guilt

III

SOCIAL BEHAVIOR

all stem from our assessment of the approval or disapproval that we receive from others.

We live in a social environment and few, if any, of our experiences and activities lack social significance. It is perhaps not too surprising, then, that our major problems are social problems. While it is true that we form friendships, marry, and join social groups, it is also true that we discriminate against others, and participate in other forms of aggression such as riots and warfare.

We usually think of love, tenderness, and cooperation as human qualities and use the term "humane" to denote kind and considerate treatment, but human beings are equally capable of behaving inhumanely. The Nazi atrocities during World War II are cited as some kind of social insanity, but what of the "normal" atrocities of every war? Are the techniques of the Spanish Inquisition no longer in vogue? Although human cruelty and brutality are much easier to recognize at a distance—especially a distance in time—there is considerable evidence that many of us in all periods of history, including this one, have been specialists in inhumanity.

Unfortunately many "normal" human beings are perfectly capable of such behavior. In one study, for example, Milgram [1965] asked subjects to deliver electric shocks to others who were engaged in learning lists of nonsense syllables; they were to administer a shock every time an error was made. The person delivering the shock could not see the person receiving the shock, but saw only a red light indicating an error. There was in fact no other person learning nonsense syllables, but most of the subjects believed that there was, and they readily responded to the instructions to increase the shock level with each error. A dial indicated shock levels which ranged from "mild" through "painful" to "lethal." Almost one-third of the subjects administered shock levels which were at the lethal point on the dial. In a subsequent study the subjects were allowed to "hear" the responses of the punished subject. Of course, since there were no punished subjects, the responses were fake. As the level of the shock increased the subjects heard protests, followed by cries of pain and pounding on the wall, followed by screams, followed by silence (at "lethal" levels of shock). Even with these conditions a large percentage of the subjects complied with the experimenter's request to shock at the lethal level.

The experimenter who conducted these studies has been harshly criticized on ethical grounds. While concern for the well-being of individuals subjected to deceitful information and instruction is certainly justified, it is likely that much of the criticism stemmed from the nature of the findings. The degree of willingness of the subjects to comply with the requests made in these studies is not a particularly pleasant observation. The findings, however disturbing, may help us to understand why atrocities are so common. We need to know why it is that over 50 million human beings were killed by other human beings in the years between 1820 and 1945. We also need to know why we routinely subject each other to less lethal forms of punishment.

There are many approaches to the study of social behavior, and each contributes from a different perspective to our overall understanding. The Arab-Israeli conflict, for example, is rooted in history. It is maintained by cultural, economic, and political factors. However, the hatred and distrust that maintain it are expressed by *individual* Arabs and Jews and arise from individual attitudes and prejudices. Historical events have a bearing on behavior through their influences on the attitudes and resulting actions of individuals. The psychologist attempts to understand social interaction by examining the characteristics of individuals. From this perspective there are no special principles of social psychology. We assume that understanding what we are like as individual human beings will provide an understanding of the bases of social accord and social discord.

The social behavior of a cat is different from that of a dog, a horse, or a lion. In fact the social behavior of a species is as unique as its physical appearance. No amount of training can "socialize" a dog into behaving like a cat or a horse like a lion. So too is man's social behavior constrained by his biological characteristics. In fact much human social behavior makes sense only in terms of man's biological heritage [Cooper and McGaugh, 1963, p. 8] :

> The animal kingdom is composed of over a million species, of which man is but one. The underlying process common to all species is adaptation, and adaptation is accomplished in countless unique ways. In effecting adaptation, each species uses specialized processes and equipment, and the efficiency of its adaptation depends upon this interplay with the total environment. A truly comprehensive view of social man is possible only if we are willing to stand back and view him as part of this vast picture of adaptation struggle. While it is true that we cannot hope to understand man's social-psychological nature unless we study him in his own right and at his own level, it is also true that a healthy appreciation for the comparative perspective will assist immeasurably in this attempt.

Biological characteristics account both for man's similarities to other animals and for his differences from other animals. For example, human social behavior depends heavily on language and the use of abstract symbols. This is possible only because of the evolution of specialized neural processes. Because of this we differ greatly from most animals, including the great apes. In many ways, however, our social behavior is like that of other animals and is subject to the same influences. As we saw in Chapters 1 and 2, the development of love is similar in monkeys and men. Our motives are also similar in many ways to those of other animals. We want food, water, sex—and the fulfillment of these wants usually involves social behavior. In fact the parallels between infrahuman and human social behavior are sometimes striking. Although female chimpanzees ordinarily do not permit mating unless they are in the estrous period of their hormonal cycle, they have been observed to permit mating by male chimpanzees in order to get food away from the male. Apparently the "oldest profession" was practiced long before man appeared on the scene.

Since we are in constant danger of annihilating the human species through warfare, we have an urgent need to know why it is that we have such difficulty getting along with other residents of this planet. A number of theorists contend that the roots of human aggression also lie in man's biological heritage—that in order to survive man evolved the physiological processes which were essential for living under dangerous and hostile conditions. Thus, much of our physiological machinery is *atavistic*. Our emotional responses of fear and anger evolved under conditions in which living depended on the ability to flee or fight. With the advent of civilization there is less need for such emotional responses, but having evolved, they remain with us.

Some writers consider the changing social conditions brought about by the rapidly increasing population density particularly significant. In his book *The Territorial Imperative* Ardrey [1966] argues that man, like other animals, is a territorial animal, and that the biologically based imperative to acquire and defend territory accounts for much of man's aggressive behavior [Ardrey, 1966, p. 236]:

The territorial imperative is as blind as a cave fish, as consuming as a furnace, and it commands beyond logic, opposes all reason, suborns all moralities, strives for no goal more sublime than survival. [One must bear in mind] that the territorial principle motivates all of the human species . . . whether we approve or disapprove, whether we like it or we do not, it is a power as much an ally of our enemies as it is of ourselves and our friends. The principle cause of modern warfare arises from the failure of an intruding power correctly to estimate the defensive resources of their territorial defender.

A related view is expressed by Morris. In *The Human Zoo* [1969], he argues that many of the problems of modern man are caused by his inability to adapt to the requirements of urban living. The conditions under which man evolved as a species differ dramatically from those under which he must now live; hence his predicament is much like that of animals in zoos [Morris, 1969, p. 8]:

Under normal conditions, in their natural habitats, wild animals do not mutilate themselves, masturbate, attack their offspring, develop stomach ulcers, become fetishes, suffer from obesity, form homosexual-paired bonds, or commit murder. Among human city dwellers, needless to say, all of these things occur. Does this, then, reveal the basic differences between the human species and other animals? At first glance it seems to do so. But this is deceptive. Other animals do behave in these ways under certain circumstances, namely when they are confined in the unnatural conditions of captivity. The zoo animal in a cage exhibits all these abnormalities that we know so well from our human companions. Clearly, then, the city is not a concrete jungle, it is a human zoo.

Although these speculations are perhaps somewhat simplistic, they are important points to bear in mind as we examine human social behavior. Such views were in fact presaged by Darwin's conviction that emotional reactions evolved in response to environmental requirements. In his book *The Expression of the Emotions in Man and Animals* [1872] he presented evidence that our emotional responses, as evidenced, for example, in facial expressions, are similar to those observed in lower animals (see Fig. 4-1). The circumstances which cause our hair to "stand on end" (piloerection) are not terribly different from those that cause hair erection in dogs and cats.

Our emotional states are accompanied by obvious physiological responses such as blushing, as well as by other, more subtle responses. We do not ordinarily reveal our social attitudes, especially our dislikes of others, by attacking behavior. However, we do respond emotionally to others, and these responses are accompanied by subtle physiological changes. Emotional responses are accompanied by a lowering of the resistance of the skin to a weak electric current. This change in resistance

86

FIGURE 4–1   *Facial expressions of animals, drawn from life by Mr. Wood [Darwin, 1872].*

is termed the *galvanic skin response* (*GSR*). Cooper [1959] has shown that social attitudes can be assessed by measuring the GSR. Prejudice is accompanied by emotional responses, and hence under some conditions the GSR can be used to measure prejudice. In one experiment subjects were read four brief statements, a derogatory and a complimentary statement about a group that they disliked and a derogatory and a complimentary statement about a group that they liked. For example, one of the derogatory statements was, "People can be divided into two groups, the good and the bad. Close to the bottom of the list are the. . . . They certainly can be said to have caused more trouble for humanity than they are worth." One of the complimentary statements was, "The world over, no single group of people has done as much for us, for our civilization, as the. . . . The world will undoubtedly come to recognize them as honest, wise and completely unselfish." As the statements were read, the subjects were instructed to think about them, but not to respond verbally to them. Similar statements were also made about groups toward which the subjects had neutral attitudes. The results were striking. In 19 of 20 subjects the GSR was greater for complimentary statements about the disliked groups than for complimentary statements about neutral groups. Moreover, in 14 out of 20 cases the GSR was greater for derogatory statements about liked groups than for derogatory statements about neutral groups. In another study Cooper used

87

comparable procedures to measure attitudes toward various ethnic groups. The GSRs were found to correlate highly with other measures of attitude obtained from conventional attitude tests—that is, tests which rely on the subject's verbal statements about his feelings regarding different ethnic groups.

Clearly our attitudes toward other individuals are accompanied by strong emotional responses. We do not merely think about others; we "feel" about them as well. Social behavior is laden with emotion. This knowledge can help us understand why it is that social interaction can be so complex. Apparently it is not possible for us to be unemotional about other human beings, or groups of human beings, that we like or dislike—and the biological aspects of prejudice are as real and important as the biological aspects of hunger and thirst.

Our emotional responses can also *influence* our attitudes toward social objects. For example, in an interesting study (Valins, 1966), male subjects viewed slides of attractive seminude females and were given evidence that they reacted emotionally to some of the slides. This was done by allowing the subjects to hear heartbeats which they thought were their own. It was found that the slides accompanied by (false) heart-rate increases were better liked than those where no change occurred. This preference was still evident on a second test a month later. This result indicates that our emotional reaction to something, even if extraneous, influences our evaluation of it. In a later experiment [Valins and Ray, 1967] it was shown that people could be made to respond less fearfully to a frightening object if they were led to believe (falsely) that they had *not* reacted emotionally to cues of it. These studies clearly show that emotional behavior is complex. We still do not know very much about the biological bases of social behavior, but what we do know indicates that a biological approach may be fruitful.

SOCIALIZATION

Socialization refers to all the processes which help to mold the individual so that his behavior is acceptable to the society in which he lives. Thus socialization is the process of making socially acceptable human beings—a process that starts on the day of birth and continues until the day of death. Socialization is continuous throughout life, because as we age the socializing influences change and the behavior and attitudes that are expected of us change. As Bugelski has commented [1956, p. 1]:

From infancy on, the to-be-civilized human being is subjected to a training process calculated to make him an acceptable member of society. He is taught where and when to sleep, eat, wash behind the ears, read, write, and calculate, to earn his living, and to grow old gracefully, or die nobly, depending upon how the great divisions of society are getting along with each other at the time.

Socializing influences are, of course, not always explicit. In fact most often they are not. The human mother is about as knowledgeable as the monkey mother concerning the way in which her treatment of an infant will influence the child's later behavior. Furthermore, explicit socialization techniques are not always effective. The parent knows much less than the animal trainer about how behavior can be shaped through rewards and punishments. Obviously socialization occurs. We are moderately—and only moderately—successful in developing the young of our society into effective members. Much of the success is clearly accidental. We do not yet know what makes good parents and teachers. Beyond that we exercise little formal control over major sources of socializing influences such as that provided by peers and by communications media. The television set constantly tells our children (and us) what they are to think, want, and expect and provides examples of behavior which can be used to obtain what is wanted. The young of our society have learned through television that, contrary to what parents may say, aggression is commonplace, accepted, and often rewarded and that killing is acceptable under some, if not most, circumstances. Furthermore, there is a great discrepancy between what we say we are like and what we actually appear to be, as reflected on the television screen. These messages, however subtle, are not lost in the young of our society (see Chapter 3).

Thus, although socialization may begin at home, where it is well under the control of parents, particularly the mother, it is not at all clear that influences from the home constitute the *major* socializing influence. One can become socialized into a society without conforming to the behavior expected by one's elders, a fact which constitutes a major source of anxiety for the adults in our society. Parents want their children to "turn out like us," and then become distressed when they do not. Changes in a society's values are most readily reflected in the young and in a rapidly and continually changing society discrepancies between the attitudes and values of parents and those of their children are likely to be great.

DEVELOPMENT OF ATTITUDES

It is at home that children first develop their attitudes toward different aspects of their worlds. However, as Krech and Crutchfield have pointed out [1948, pp. 181–182]:

> [To say that parents are important in shaping attitudes] is not equivalent to saying that the child will take over attitudes and beliefs ready made from the parents. The influence is possible, but whether the child will develop . . . the same beliefs as [those held by its parents] . . . depends upon the importance and meaning of that belief for the child. . . . In some instances the effect of the parents' influence can be seen to account for the rise of a belief or attitude that is in opposition to the parents' belief.

Studies of the attitudes of college students and their parents indicate a moderate but significant relationship; generally the correlation coefficients range from about +.30 to +.60, depending on the attitude being measured [Hirschberg and Gilliland, 1942]. There is evidence that the degree of similarity varies with the attitude of the child toward the parent. College students who have unfavorable attitudes toward their parents tend to have social ideologies which are quite dissimilar from those of their parents, while in students who have favorable attitudes toward their parents there tends to be less of a discrepancy [Cooper and Blair, 1959].

It seems clear that exposure to parental attitudes is not enough to ensure their adoption by children. If the attitudes of the parents are consistent with those of other segments of society, the likelihood of similarity in attitude is increased.

Some of our attitudes are based on firsthand information. This fact is so obvious that we may be led to believe that information provides the basis for all of our attitudes. We may like ice cream because we know how it tastes, and we may like certain sports because we find them exciting. Social attitudes, however, are often based on the prevailing attitudes of the society. The attitudes adopted are very likely to be those to which the child is exposed. For example, actual contact with minority-group members is unnecessary for the development of unfavorable attitudes toward such groups. Radke and Sutherland [1949] reported highly developed prejudices toward Jews and blacks in Midwestern children who had had little if any contact with members of either group. In fact Rosenblith [1949] found that in a region of South Dakota where there were no Jews or blacks at all, prejudice scores were higher than those in areas where there had been extensive contact with both.

Experience with minority-group members is not any guarantee of favorable attitudes toward them. In a classical study of the development of attitudes toward blacks Horowitz [1936] found that boys from New York City were about as prejudiced as were children from Southern states. On the basis of these findings Horowitz concluded [1936, pp. 34–35]:

It has been found necessary to contradict many of the oft-repeated cliches current in the discussion of the race problem. Young children [are] not devoid of prejudice; contact with a "nice" Negro is not a universal panacea; living as neighbors, going to a common school, were found to be insufficient; Northern children were found to differ very, very slightly from Southern children. It seems that attitudes toward Negroes are now chiefly determined not by contact with Negroes, but by contact with the prevalent attitude toward Negroes.

Of course much has happened in our society since 1936. Nevertheless the general conclusion is probably still valid. This finding clearly poses a formidable problem for our democratic society. Since most of our social problems grow out of our emotion-laden attitudes, it is highly unlikely that the problems can be dealt with adequately without changing the attitudes which have led to their development and continue to support them.

Most of our social conflicts grow out of dislike for other humans, but fortunately we also sometimes like our fellow man. What factors influence social attraction among individuals? What is it that attracts us to one person rather than another? It is only recently that social psychologists have begun systematically to search for variables affecting the social attraction of one individual to another.

SOCIAL ATTRACTION

For the young of many species affiliation with mature species members, especially the mother, is virtually a prerequisite for survival. Imprinting provides a mechanism by which chicks or ducklings are directed to exhibit affiliative behavior toward the imprinted object, and the clinging responses of infant monkeys help to ensure that they will remain within the protective grasp of their mothers for appropriate periods of time. There is strong evidence that for most species, including humans, unlearned response patterns account for much of the affiliative behavior exhibited in early postnatal development.

Through the processes of socialization affiliative responses, initially reflex in nature and directed toward the most readily available social agent, gradually come to be discriminatively directed toward specific members of the social group. In Chapter 1 and 2 we examined the processes by which the infant monkey's social behavioral repertoire expands beyond initial relationships with the mother to include the peer and subsequently the heterosexual affectional systems. By adulthood affiliative responses constitute only a fraction of an individual's social repertoire, and exhibition of such responses is, of course, no longer controlled by reflex activity alone. Animal research is enormously useful as a basic model of similar development in humans. However, the adult human organism leads an exceptionally complex social life. Although maternal love, for example, is undoubtedly a prerequisite, the basis of specific social preferences of the adult is not fully explicable in terms of such global concepts. We turn now to a consideration of some of the factors which influence the attraction of individuals to other individuals.

SIMILARITY

It is consistently found that the more similar individuals are in attitudes and interests, the more they are inclined to like each other. For example, in one study [Byrne, 1961] subjects were asked to fill out question-

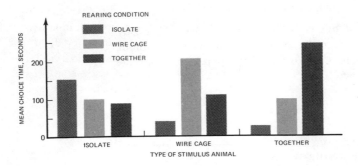

FIGURE 4–2   *Social preference and similarity of rearing condition for young rhesus monkeys [Pratt and Sackett, 1967].*

naires rating certain attitudes and interests, such as political positions and hobbies. At a later date the same subjects were asked to "score" similar questionnaires supposedly filled out by others. In reality the questionnaires had been prepared to be either very similar, slightly similar, or very dissimilar to the subjects' own ratings. The subjects were then asked to indicate how much they thought they would like the individual whose questionnaire they had scored. A surprisingly strong relationship between perceived similarity and expressed attraction was disclosed. Other studies of the attitudes and interests of good friends as opposed to casual acquaintances or people who were not friends have quite consistently shown that the more people like each other, the more similar they tend to be.

Highly similar results are found with rhesus monkeys. Pratt and Sackett [1967] found that when monkeys reared in social isolation, in individual cages, or in the presence of peers were allowed to choose among an isolate-reared, a cage-reared, or a peer-reared stimulus monkey, isolates tended to prefer isolates, cage-reared preferred cage-reared, and peer-reared preferred peer-reared stimulus animals (see Fig. 4-2). In this study similarity of rearing experience was clearly the primary factor in social preference. In another monkey study [Suomi et al., 1970] normal monkeys and monkeys with frontal lobectomies (with the frontal lobes of their brains surgically removed) were allowed to choose between normal and lobectomized monkeys either of their own or of the opposite sex (see Fig. 4-3). In choosing between animals of their own sex the lobectomized animals displayed no preference, while the controls preferred lobectomized animals. However, in choosing between animals of the opposite sex the lobectomized animals chose lobectomized partners, while the normal animals chose to be with normal animals. In their discussion of these findings Suomi et al. commented [1970, p. 452]:

[As the monkeys] . . . had not personally observed the operations performed on their peers we must assume that their choices were made on the basis of observed behavioral differences. . . . The specific nature of these differences . . . are known only to the monkey subjects themselves.

It is obvious that monkeys are quite capable of discriminating among monkeys that look the same to human observers and, moreover, they apparently have good reasons for making the choices they do. The suggestion is that in some way monkeys are probably more knowledgeable observers of behavior than are humans—which need not be taken as a blow to an experimenter's pride. After all, they have had a bit more experience than we.

The assumption that likes attract likes underlies the computer-dating industry. Of course, as even monkeys probably know, perceived similarity is not all there is to social attraction, particularly the attraction to someone of the opposite sex. This fact was made abundantly clear by the results of one study using computerized matching of chance partners. Walster et al. [1966] arranged a computerized-date dance with partners matched for varying degrees of similarity. Partner attraction was measured both during the dance and in several followup questioning periods over the succeeding three months. The experimenters found, perhaps to their surprise, that perceived similarity to one's date had very little effect on attraction. They did discover, perhaps to no one else's surprise, that physical attractiveness of one's date had a whopping effect on how much the date was liked.

Of course the factors that influence attraction to dance partners are likely to be different from those influencing social attraction under different conditions.

Often we want to associate with others who have shared or will share similar experiences. People who are generally happy may not wish to relate to those who are depressed, and people in a festive mood usually choose not to mingle with those who are not. There is experimental support for the adage that misery loves company. In one study, for example, Schachter [1959] told female college students either that they would be subjected to mildly painful electric shock or that they would engage in a nonstressful task. All subjects were then told that they would

FIGURE 4–3   *Social choices of normal monkeys and monkeys with frontal lobectomies [Suomi et al., 1970].*

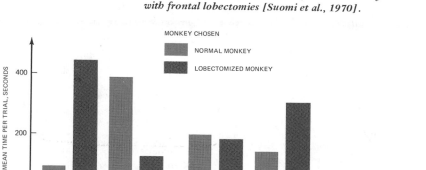

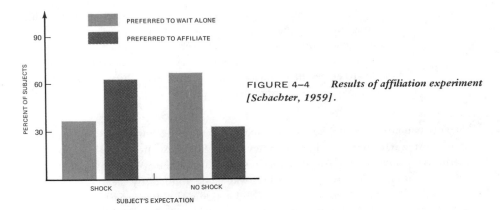

FIGURE 4-4   *Results of affiliation experiment [Schachter, 1959].*

have to wait a few minutes before the experiment began and were asked to indicate whether they preferred to wait alone or in a room with other people. After the subjects had indicated their preferences the experiment was finished. Subjects who thought they were about to be shocked showed a greater tendency to wait with others than did subjects who did not anticipate shock (see Fig. 4-4).

In another study, also of female college students, all subjects were told that the experiment in which they were to participate involved electric shock, that they would have to wait before the experiment began, and that they had the choice of waiting either alone or in a room with others. If they chose to wait with others, they could elect to wait with other subjects who were waiting for the experiment to begin, with other subjects who had supposedly already been through the experiment, or with a group of people not involved in the experiment. Among those subjects who indicated a preference for waiting with others rather than alone, the overwhelming choice was to wait with people who were themselves about to be shocked. Apparently "miserable people seek out miserable company." In a third study Schachter found that if subjects were moderately anxious about the impending shock experiment, they preferred to affiliate with others who were also moderately anxious rather than those who were either extremely anxious or scarcely anxious at all. The adage can be further refined—"miserable people seek out miserable company who are about as miserable as themselves."

Why is it that people experiencing stress or anxiety tend to seek out others who share their predicament? To begin with, this finding is not limited to affiliation arising from emotive factors. Several studies have demonstrated that under appropriate conditions people will also choose to affiliate with others whose beliefs or abilities are similar. These observations have led to general theoretical interpretation termed *social comparison* [Festinger, 1954; Schachter, 1959; Latané, 1966]. According to this view, there exists in humans a tendency or need to evaluate one's abilities, opinions, and emotions, and in the absence of objective, nonsocial standards for comparison, people evaluate these elements by comparison with others. Hence a person who is unsure of his ability to perform a given task when objective standards are unavail-

94

able tends to affiliate with others who perform the same task. A person who is unsure of a belief or the validity of his emotional feelings will seek out others in a similar position in order to compare and evaluate his own beliefs and feelings.

According to social-comparison theory, people will select those whom they perceive to be relatively similar in the quality for which comparison is desired. For example, a moderately competent chess player will seek a partner who is neither a novice nor an expert, but rather someone in his own class. A member of the "silent majority" will tend to affiliate with politically moderate conservatives rather than with radicals or reactionaries. A patient awaiting minor surgery will prefer the company of others expecting an uncomplicated operation rather than the company of those merely taking a physical examination or those about to receive a heart transplant. Thus there is considerable evidence that birds of a feather do flock together.

Although perceived similarity is apparently a major factor in the way we choose to relate to others, another variable hypothesized to affect social attraction is *complementarity* of personality, or attraction of opposites. We might intuitively expect a sadist and a masochist to get along better together than either two sadists or two masochists, but there have been few studies of the variability of complementarity, probably because it is a difficult measure to define. There is some empirical support for the hypothesis. Married couples who are complementary in personality sometimes have stronger relationships than those who are not. However, the role of complementarity in influencing interpersonal relations has not yet been adequately studied.

PHYSICAL PROXIMITY

Common sense tells us that two people will rarely fall in love if they live 3000 miles apart and have never come into any sort of contact. However, the significance of functional proximity as a factor in determining the strength of social attraction is not so obvious. It may surprise some to learn that well over one-third of the urban marriages in this country are of couples who grew up less than six city blocks from each other.

One of the first empirical studies to emphasize the importance of proximity was a field experiment conducted by Festinger et al. [1950] in a newly constructed housing development, part of which consisted of two-story rectangular units with stairways at both ends. The experimenters collected data, by means of both questionnaires and personal interviews, from most families in the entire housing development when the development was first opened and over a period of many months afterward. On each of the surveys families were asked to name their best friends in the housing development. After an initial period of fluctuation, some consistent findings emerged. First, the nearer families lived to each other, the more likely they were to become close friends. Within each unit family 1 was more likely to be friends with

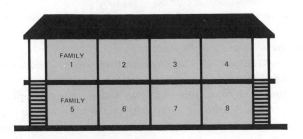

FIGURE 4–5   *Arrangement of apartments in two-story dwelling units [Festinger et al., 1963].*

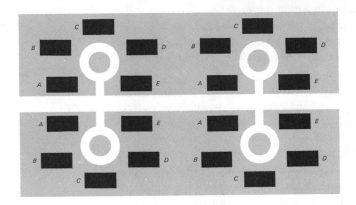

FIGURE 4–6   *Ground plan of housing units.*

family 2 than with family 3, and more likely to be friends with family 3 than with family 4 (see Fig. 4-5). Moreover, families living next to the staircase were significantly more popular (reported to be friends) than families living midway down the halls. Since all people on the second floor had to use the stairs to reach their doors, they would be more likely to meet people living next to the stairs than those living midway down the hall. The setup of apartment buildings is shown in Fig. 4-6. Families living in adjacent buildings were more likely to be friends than families living several buildings apart. Also, people living in units A and E tended to be more popular than those living in other units. Clearly functional proximity is an important determinant of social attraction. Similar findings were obtained in a study of the development of dormitory social relationships over a period of a year [Newcomb, 1961]. Proximity was positively related to degree of social attraction between individuals; roommates and floormates were more likely to be close friends than members of different dormitories.

Why should functional proximity affect social attraction? One possibility is that the closer two people are physically, the greater the opportunity for interaction. It is easier for people to become familiar with one another when they are physically close than when they are physically far apart. It is encouraging to know that people who live in relatively close proximity tend to like each other; it means that human relations are not altogether hopeless. However, whether familiarity will

96

lead to liking or breed contempt depends on many factors, including the nature of the people who come into contact with each other as well as the nature of the relationship. The ghetto resident who is constantly harassed by law-enforcement agencies will not necessarily come to like them. The high incidence of divorce also testifies to the fact that proximity and familiarity do not guarantee social attraction.

Another characteristic which influences social attraction is *rewardingness*. Simply put, people tend to like those who reward them with love or praise. Rewardingness is not an easy characteristic to assess, since what is rewarding to one person may be aversive to another. However, research has shown that people tend to like those who like them. Subjects report that it is usually reinforcing to be liked, and that they tend to prefer the pleasant company of those who like them rather than be with people who dislike them.

It has also been demonstrated that praise or criticism from an individual affects attraction to him. Subjects were placed in a group-discussion situation where they periodically received evaluations from group members concerning their contributions to the discussion [Aronson and Lindner, 1965]. The other members were paid confederates, and the evaluations were experimentally controlled. Some confederates consistently gave the subjects "positive feedback"; these people were generally liked by the subjects. Others consistently gave negative feedback, and not surprisingly, they were not as well liked. Other confederates played a traitor role; initially they gave the subject positive feedback and then switched and gave negative feedback. These people were the least liked of all. Finally, some confederates initially gave negative feedback and then switched to support of the subjects. These people were the most liked of all. Apparently being able to win over a dubious skeptic is more rewarding than receiving positive support from the outset.

Some cautions regarding the findings of the research on social attraction should be expressed at this point. First, most studies have attempted to hold all variables constant except the one in question. Thus we cannot conclude that rewardingness, for example, tends to enhance attraction; we can say only that all other variables being held constant, rewardingness tends to enhance attraction. This is an important distinction. Most of the research in this area has been on specific factors that may affect attraction. Relatively little work has been directed toward untying the complex interactions which certainly exist among the several variables. Second, much of the research is based on contrived and perhaps trivial social situations. We have much to learn about the bases of the formation and dissolution of attraction between people in real life situations.

97

CONFORMITY

Conformity is one of the most obvious and basic characteristics of social behavior. In the most general sense *conformity* refers to a person's tendency to behave like those around him. In a more specific sense it is the adherence to a norm of behavior set by a group [Jones and Gerard, 1967] —conformity to fashion tastes, conformity to styles of life, or even conformity to the act of being a nonconformist. The classical laboratory demonstration of conformity was performed by Asch [1951]. He presented a group of eight people with a slide of a line segment. He then flashed a second slide containing three line segments and asked each one to report verbally, in front of the rest, which of the three lines was most similar to the first line segment (see Fig. 4-7). After a specified number of the experimental trials each of the first seven people, who were in fact confederates rather than subjects, reported line *A* to be correct rather than the obviously correct line *B*. Asch was really interested in how the eighth person, the only real subject in the experiment, would respond. It is apparent from the figure that the subject was faced with an easy perceptual task. On earlier trials the consensus of the group had been in his favor, but suddenly everyone else was reporting the "wrong" answer. Asch found that in all such tests the subject showed a great deal of uneasiness, agitation, and hesitation before answering. Moreover, in approximately one-third of the cases the subject also reported the wrong answer as correct, in conformity to the answer given by the group.

This type of experiment has been replicated many times with exceptionally consistent findings. In some situations there appears to exist overwhelming pressure to conform to group norms, even when such conformity is clearly at odds with the subjects' own perceptions, attitudes, or beliefs. Further studies have isolated specific variables which tend to enhance or decrease the conformity behavior of subjects placed in the above situation. First, the size of the consensus group seems to be a factor. In the experiment above four confederates were needed to produce the maximum degree of subject conformity. With fewer than four the subjects were less likely to report conforming views, and more confederates than the "optimal four" did not result in sizable differences in either direction.

A second variable is status of the confederates. Subjects are more likely to conform when they perceive the group members to be equal to or higher than themselves in social status and less likely to conform when they view the group as below them in status. A related variable is "expertness" of the confederates. If, instead of line segments, the stimulus objects are mathematical equations, a subject is more likely to conform to the judgment of confederates purported to be math instructors than those introduced as art majors—that is, he is more likely to conform to the opinion of experts or authorities than to the opinion of novices.

A third variable found to affect tendency to conform is the difficulty of objective discrimination. For example, in the two situations

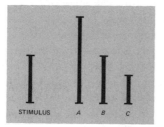

FIGURE 4–7 *Typical stimuli in Asch experimental paradigm.*

FIGURE 4–8 *Two Asch situations differing in difficulty of objective discrimination.*

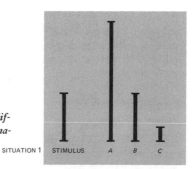

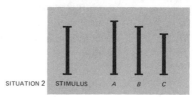

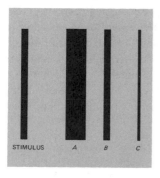

FIGURE 4–9 *Another typical Asch paradigm.*

shown in Fig. 4-8, segment *B* is the correct choice in both cases. However, subjects show a much greater rate of conformity to the group's choice of *A* in situation 2, where the distinction between *A* and *B* is much less extreme.

Other studies have focused on variables which decrease conformity in the Asch experimental situation. By far the greatest influence is the presence of other nonconforming members in the group. For example, when one of the seven confederates deviates from the group's decision, subject rate of conformity drops significantly. If two confederates deviate from the group's incorrect answer, virtually no subjects will go along with the incorrect evaluation. The presence of another nonconformer exerts a powerful effect even if he is dissenting to the correct answer to a problem. In Fig. 4-9, where the correct answer is obviously *B*, if all seven confederates choose *A*, then, as we have seen, the subject also tends to choose *A*. If one of the confederates chooses *B*, the subject is much more likely to make the correct choice of *B*. However, if one of the confederates chooses *C*, which is incorrect but also a nonconforming answer, the subject is less likely to agree with either the group or the dissenter and will tend to choose *B*.

99

It is interesting to note how the mere order of answering affects the manner in which subjects caught in an Asch situation evaluate the various confederates. For example, in a group of three confederates and one subject, where the subject is always the third person to answer and his judgment always differs from that of the three confederates, the subject consistently tends to evaluate the other members of his group as follows. The first confederate to choose a stimulus on each trial is seen as powerful but incompetent; he has the power to make the "right" choice and he invariably fails. The second person to answer is seen as a sniveling lackey, an insignificant yes-man who is too timid to disagree. The subject answers third. His evaluation of the fourth person, who follows him, is usually extremely negative. This person is a traitor who might have supported him but instead chose to lie and support the first two.

Some of the most extensive research on yielding to group pressure was conducted by Tuddenham. The stimuli were not restricted to simple line drawings, but dealt with various kinds of historical, geographic, economic, and social judgments. Many of the examples presented were sufficiently extreme or bizarre that there was relatively little subject conformity. Nevertheless, there were a few extreme yielders, individuals who made affirmative responses to extreme statements when they had to respond in the presence of others who had made affirmative responses. If they had yielded completely to all such statements, this is what they would have been agreeing to (I indicates an information item and O indicates an opinion item) [Tuddenham and MacBride, 1959, p. 260]:

[The United States] is largely populated by old people, 60 to 70% being over 65 years of age (I-1). These oldsters must almost all be women since male babies have a life expectancy of only 25 years (I-2). Though outlived by women, men tower over them in height, being 8 or 9 inches taller, on the average (I-4). This society is obviously preoccupied with eating, averaging six meals per day (I-5), this perhaps accounting for their agreement with the assertion, "I never seem to get hungry" (O-9). Americans waste little time on sleep, averaging only 4 to 5 hours a night (I-3), a pattern perhaps not unrelated to the statement that the average family includes 5 or 6 children (I-9). Nevertheless, there is no overpopulation problem, since the USA stretches 6,000 miles from San Francisco to New York (I-6). Although the economy is booming with an average wage of $5/hour (I-7), rather negative and dysphoric attitudes characterize the group, as expressed in their solidly rejecting the proposition, "any man who is able and willing to work hard has a good chance of succeeding" (O-3), and in agreeing with such statements as, "Most people would be better off if they never went to school at all" (O-5), "there is no use in doing things for such people, they don't appreciate it" (O-6), and "I cannot do anything well" (O-10). Such is the weird and wonderful picture of the world and of themselves allegedly entertained by the "others in the group."

groups it is often the case that two leaders will emerge, one of each typ[e].

We know very little about how leadership is developed, but som[e] recent studies suggest that its origins may be far from noble. In a classica[l] experiment by Bavelas et al. [1965] spontaneously formed groups were given a variety of problems to discuss, ranging from the utility of free-trade legislation to whether the United States should recognize Red China. After each 20-minute discussion session all members were asked to indicate which person they thought was the leader. Within a few sessions a clear consensus usually emerged, with the leader typically being the person who talked the most. This is not a unique finding.

The group members were then told that their meetings would be monitored by "experts" who would evaluate each person's participation in the discussion. A set of lights was placed in front of each subject, hidden from the view of the other group members, and the subjects were told that every time they made a positive contribution the green light would go on and every time they detracted from the general train of the discussion the red light would go on. The goal was to make the person who had received the lowest leadership rating the leader of the group. Each time this person spoke, no matter what he said, he received a green light, while the other members consistently received red lights for their contributions to the discussions. Despite the fact that each group member would see only his own set of lights, eventually the previously low man was doing most of the talking, and when the next leadership pool was taken, he emerged as the group leader. Moreover, in the succeeding discussion, carried out in the absence of "experts" and lights, the new leader remained the chief, both in the amount he talked and in the postdiscussion opinion poll. Apparently it took only a green light to make a leader of a lackey.

Of course this experiment does not refute the possibility that there are born leaders. However, it does illustrate that many leaders can be made, quickly and relatively easily, out of virtual nobodies if conditions are proper. Anyone familiar with our national politics should view this finding with little surprise.

CHANGING SOCIAL BEHAVIOR

A person's attitudes reflect the way he perceives things in the world about him—other people, objects, opinions, beliefs. People are not born with the attitudes they hold. Rather, attitudes are learned, and as such are undoubtedly affected by subsequent experiences. Thus they may change with time, some more easily and more readily than others. The process of attitude acquisition and the mechanism of attitude change represent perhaps the most thoroughly investigated research topic within social psychology.

The practical implications of determining the mechanisms of attitude change are obvious. Selling a product, electing a political

Obviously some degree of conformity or compliance with the group is essential for adapting to a social environment. It is also probably useful in many circumstances, since the beliefs and opinions of one's peers are not always outrageous. Nevertheless, it is fortunate that most of us do not yield so readily to group pressure that we would share such extreme beliefs as those above.

GROUP PROCESSES AND LEADERSHIP

Loosely speaking, a *group* can be described as a collection of individuals. However, groups can have such an enormous range of size and purpose that it is difficult to make useful generalizations about group processes. Hence some sort of classification of groups is necessary. We can distinguish between *nominal groups*, collections of individuals who function relatively independently of each other, and *true groups*, collections of individuals who interact with each other. The activities of nominal groups are seldom analyzed in the same terms as those of true groups which have some express purpose or function. Among true groups, we can further differentiate between *task-oriented groups* and *process-oriented groups*, at least at a hypothetical level. The goals of task-oriented groups are generally expressed in terms of work output. Members of an assembly-line unit or a committee to prepare a report on a specific project would qualify as a task-oriented group. A typical example of a process-oriented group would be a collection of individuals engaged in sensitivity training. Here the expressed goal is not that of producing an objective output, but rather of achieving a certain level of interpersonal understanding. In many cases, of course, the distinction is arbitrary. It is easy to conceive of a collection of individuals who would qualify equally well as a task or process group.

There are numerous other ways of classifying groups—leaderless groups versus those with a designated leader, structured versus unstructured groups, permanent versus temporary groups, static versus dynamic groups. The primary reason for making such distinctions is that variables found to affect performance in one type of group may have very different effects on another type of group.

COMPETITIVE VERSUS COOPERATIVE GROUPS

Deutsch [1949] formed artificial groups of two types, each consisting of five subjects. Each group met for 3 hours per week for six weeks in order to fulfill the requirements for an introductory class in psychology. The cooperative groups were told that they were to function as a team, and that the course grade of each person in the group would depend on the merits of the group solution. The competitive groups were given the same tasks, but they were told that each of the five members would receive a different grade, depending on his relative contribution to the group's solutions of the various problems. The cooperative groups were generally more productive. They solved simple tasks more rapidly than the competitive groups and produced more realistic recommendations for solution of complex tasks. There was more communication in the cooperative groups, and the members of these groups reported having fewer difficulties understanding each other. One could hardly paint a rosier picture for the spirit of cooperation.

Related experiments, termed the "bastard studies," indicate that the picture is less than rosy in cooperative groups if the subjects fail to cooperate [Hastorf, 1966]. Typically, subjects are assigned to groups and told they are to perform individually a menial task for which they will receive a numerical score. If each member of the group achieves a score greater than 100, the entire group will receive a substantial monetary reward; however, if even one member fails to achieve the criterion score, no one will be paid. After subjects perform the designated tasks, they are told that their scores came out as shown in Fig. 4-10. The subjects are then asked to rate how they feel about each of the other members of the group. Of course the scores are fictitious and subject D is a paid confederate, but the following results are consistently obtained. In condition 1 the typical attitude expressed toward D, the paid confederate who has failed, is one of sympathy. One would have to be really incompetent to perform so inadequately, and the poor fellow probably deserves pity rather than anything else. Condition 2 produces far different evaluations of D, the most typical being "that bastard!" (hence the study's nickname). Here D is usually seen as possessing the ability and having failed the rest of the group only through lack of motivation; thus he is perceived as the sole cause of the group's failure. Obviously the effectiveness of a cooperative group depends on the behavior of its individual members. Nevertheless, most studies indicate that individuals in cooperative groups have higher morale, discuss problems more freely, and report greater satisfaction than those in competitive groups.

GROUP ORGANIZATION

Groups can also differ in terms of the distribution of authority. For example, Lippitt [1940] studied five-person groups of fifth- and sixth-grade children with both democratic and authoritarian leaders. With the authoritarian leadership the children were restricted to group tasks and were allowed no individual goals. All assignments were doled out, and the decision of which children were to work together was arbitrarily decided. In the groups with democratic leadership the goals and purposes were explained, and the children were given a role in deciding the goals to be achieved. There were numerous differences between the two types of groups. In general the democratic situation created an atmosphere of easy communication and mutual help. The children produced better work, were less dominating and more friendly to each other, and showed fewer instances of aggressive behavior to each other.

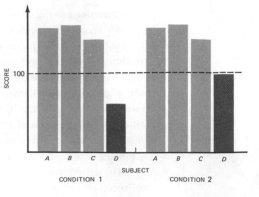

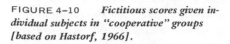

FIGURE 4–10   *Fictitious scores given individual subjects in "cooperative" groups [based on Hastorf, 1966].*

In a more elaborate study by Lewin et al. [1939] groups of boys were exposed not only to authoritarian and democratic types, but also to a laissez-faire situation in which the leader allowed virtually complete freedom, refrained from participating himself, and made few comments. As in Lippett's study, the children under authoritarian leadership were more aggressive or apathetic, and those who were apathetic tended to become aggressive when the leader left the room. It was also found that the children liked the leader better when he functioned in a democratic role than when he behaved autocratically. They even preferred laissez-faire leadership to authoritarian leadership.

Group processes are profoundly influenced by the type of leadership provided, but what differentiates an effective leader from an ineffective one? Are leaders born or made, or is this impossible to determine? Most of these questions have been asked for years, and we still do not have definitive answers. However, we do know quite a bit about leaders and leadership.

Leadership styles tend to fall into one of two classes—task oriented or process oriented—and the two types of leaders differ markedly in personality. A task-oriented leader, as the designation implies, is primarily concerned with pushing the group to create a finished product. He typically talks more than other members of the group, directs and issues orders, and executes rather than administrates. His concern is more with what is produced than with how it is produced. A process-oriented leader's *modus operandi* is almost the opposite; his primary concern lies in keeping the group intact, in maintaining group cohesiveness. He advises rather than issues orders, and listens more than he talks. He is more directed toward solving personal difficulties within the group than with meeting a production goal. Which type is better? Obviously this depends on the nature of the group being led and who is doing the evaluating. A process-oriented leader would probably not make an effective Marine Corps drill sergeant, while a task-oriented leader would be out of place in a sensitivity-training encounter group. A foreman who was a task leader might be praised by his contractor but loathed by his men; a process leader in the same position might claim the respect of his workers but not that of his employers. These two qualities are rarely found in the same individual. However, in spontaneously formed

candidate, reducing prejudice, accepting the need for population and pollution control, or educating parents and children on the effects of drugs all depend ultimately on the success of a communicator in shaping or changing the attitudes of his audience. The variables which affect the communicator's effectiveness are of interest both to those who seek a world of change and to those who would not "rock the boat."

Consider a political speaker trying to persuade an audience that the Republicans are correct on a particular issue and that the Democratic position is untenable. Will he be more successful if he presents only the Republican argument, or will he persuade more members of the audience if he includes some of the points made by the Democrats before presenting his position? Studies by Hovland et al. [1949] and Lumsdaine and Janis [1953] indicate that it depends on the political composition of the audience. The findings suggest that one-sided arguments are more effective in cementing existing attitudes, while two-sided arguments produce more change in a skeptical audience. Presentation of opposing views detracts from an argument supporting a position with which the audience agrees and may raise doubts among those who were initially in favor. However, where the audience is initially opposed a one-sided argument may be dismissed as unfair and unrealistic.

Another variable found to affect degree of attitude change is the speaker's credibility. Very simply, an audience is more receptive to arguments from a source they believe. This has been borne out both in numerous experimental situations and in public-opinion surveys indicating that a credibility gap accompanies a decrease in popularity of administrative policies. Credibility appears to be affected by both expertise of the source and coorientation of the source with the audience. Generally speaking, the greater the perceived expertise of the source, the more effective the persuasion. On the issue of a critical world food shortage people are more likely to believe the opinion of an eminent ecologist than that of an Indian mystic. A man interested in buying stocks is more likely to consult the *Wall Street Journal* than *The Militant* to determine the current economic conditions of the country. A jury will probably take the opinion of a court psychiatrist more seriously than the opinion of an eighty-year-old relative concerning the sanity of a defendent.

However, Hovland and Weiss [1951] have shown that the persuasive effects of expertise tend to diminish over time. More specifically, attitude change resulting from a source of high expertise diminishes over a period of weeks, while change resulting from a low-expertise source appears to increase (see Fig. 4-11). This *sleeper effect* has been explained on the basis that people forget the source of a persuasive communication more rapidly than they forget the content of the communication.

A second factor affecting credibility is the degree of coorientation of the source with the audience. The more a speaker appears to share the views and values of his audience, the more the audience will be influenced by what he has to say. Members of the John Birch Society are more likely to believe the arguments of Spiro T. Agnew than the arguments of Abbie Hoffman. The people of a ghetto are more likely to accept the argument

105

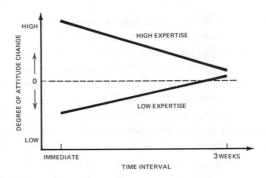

FIGURE 4–11   *The sleeper effect in the influence of communication
source over time [Hovland and Weiss, 1951].*

that a black suspect was fleeing when he was shot by police than the
police department's explanation that the officer shot in self-defense.
Drug users respect the opinion that "speed kills" more when the source
is an ex-addict than when the source is a narcotics detective. In short,
the degree to which an audience can identify with the source of information is an important factor in credibility. Being able to "tell it like it is"
can be a powerful asset in bringing about attitude change.

Closely related to the credibility of the source is the plausibility
of the argument. For example, fear-producing communication has been
found to be a relatively useful technique in effecting attitude change particularly with respect to issues such as cigarette smoking, health, pollution, and the "threat" of Communist conspiracies. Generally speaking,
the more disastrous the predicted consequences, the more likely the attitude change. However, this can be overdone. Janis and Feshback [1953]
presented grade-school children with one of three lectures stressing the
importance of oral hygiene and then measured the effectiveness of the
lectures by surveying parents on any change in the children's dental-care
habits. The lectures differed only in the amount of fear-arousing material; one discussed cavities and bad breath, the second showed some
slides of minor cavities, and the third emphasized with gory illustrative
slides the painful consequences of severe tooth decay and diseased gums.
All lectures stressed that brushing teeth up and down rather than sideways would prevent dental problems.

The results indicated that the second condition was most effective, while the third produced a "boomerang" effect and promoted less
attitude change than the mildest lecture (see Fig. 4-12). Two explanations have been offered. The first is that people avoid perception of intense fear-producing stimuli; by avoiding the message they avoid facing
the problem. This has been termed *selective exposure*. A second explanation concerns the plausibility of the proposed cure; subjects are not
likely to believe that brushing up and down rather than sideways will
make a critical difference between healthy and diseased gums, but it
might make a difference between cavities and no cavities.

The factors we have discussed concern the source and type of per-

106

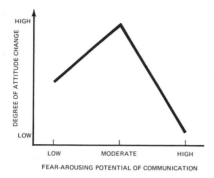

FIGURE 4–12   *The boomerang effect of fear-arousing material on attitude change [Janis and Feshback, 1953].*

suasive communication. However, another important factor is the nature of the attitude being subjected to change, especially its strength and saliency and the extent to which it is intimately tied to an entire value structure. These points are most likely to arise in matters that touch on racial prejudice or religious convictions. An attitude of this nature tends to be exceedingly resistant to change.

Obviously a complete understanding of the process of attitude change is far beyond the present state of knowledge. Despite current fears that ours is becoming a brainwashed society, controlled by those who control the media, in reality remarkably few advertising or political campaigns actually resulted in mass shifts of public opinion. It is a fortunate politician whose campaign sways more than 10 percent of the populace, and a rare new product that immediately captures the buying market from existing competitors.

We often hear the phrases "the battle for the minds of men" or "control of the mind." There is current concern in many circles that with the advance of technology we are gaining an increased ability to exercise control over the motivation and behavior of human beings. Most often the concerns are expressed as fears. There is considerable concern, for example, that manipulation of the mind through direct treatment of the brain can lead people to engage in behavior that is "against their will." Since it is possible to control the behavior of rats by delivering small electric currents to specific regions of the brain, is it also possible that such techniques will be used to control the behavior of human beings?

In some ways it is paradoxical that such questions should be asked, because from another point of view we are greatly concerned that we do not have enough control over the behavior of members of our society. Educational institutions of our society are specifically designed to help us socialize our children in specific and controlled ways. In a more forceful way, law-enforcement agencies and the armed forces are designed to exercise control over men.

The point is that the behavior of men can be rather effectively controlled by the agencies of our society, and in some cases the degree

107

SOCIAL BEHAVIOR

of control is greater than that which could be expected by placing electrodes in the brains of humans. As a matter of fact, although that particular technique has been found to be effective in the control of severe emotional outbursts in severely disturbed individuals, it would be both dangerous and uneconomical for the control of social behavior in the "normal" individual.

The major question, then, is not whether human beings have the capability of exercising control over other human beings; clearly they do, as much of the evidence presented throughout this book testifies. The concern appears to center on the nature of the control and who is to be in control—the age-old question of *qui custodiat custodes*. We seem to be less concerned about controls that are exerted through normal communications processes than we are about controls which involve direct interference with brain function. Even here, however, we are selective. We object vigorously to the use of certain drugs, particularly those which are preferred by the youth of our culture, but we readily accept the use of other drugs, such as alcohol, nicotine, and caffeine, which are regularly used by the adults in our society. Drugs, of course, provide direct and effective control of behavior through their influence on brain function.

The most important problem and the greatest source of concern is that of who has the power to control the behavior of others. It seems as though most of us are worried about the possibility that others will gain too much control over our behavior. At the same time we seem to deplore the fact that we have too little control over the behavior of others. The effective means of control—that is, of influencing the behavior of other human beings—are available to us all. The most effective means are, of course, our ability to influence the behavior of others through our social relations with them. The fact is that we know more about how to influence behavior than we do about the bases of the influence. Adolf Hitler was able to exercise control over millions of people without a complete understanding of the nature of social influence. However, if we are to increase our ability to get along with other human beings with whom we share our planet, we need to make every effort to increase our understanding of human social behavior.

SUMMARY

Biological characteristics account for both the similarities and the differences in the social behavior of each species. Although human social interaction depends heavily on man's capacity for language, it is in many respects based on emotional states similar to those observed in many other animals. Emotional responses such as those which accompany prejudice can sometimes be measured by changes in galvanic skin response.

Socialization, the process of making socially acceptable human beings, begins at birth under the control of parents, but the home is not necessarily the major socializing force. Although attitudes and beliefs may be fashioned initially by parental example, they are altered and maintained by such other influences as the child's attitude toward his parents, the influences of peers, and the prevailing attitudes of society. Thus mere exposure to parental attitudes or to firsthand information does not ensure the adoption of particular attitudes. Through the process of socialization affiliative responses, which begin as reflexes, become directed toward specific members of the social group.

Particularly important in determining social attraction between individuals is the variable of perceived similarity. For example, the more similar two individuals are in attitudes and beliefs, the more likely they are to affiliate. This is true in many circumstances for nonhuman as well as human primates. Complementarity of personality has also been considered relevant in determining social attraction among individuals, although empirical support is not as strong for this influence. Proximity, familiarity, and rewardingness have all been found to affect social affiliation and attraction, but much is yet to be learned about social attraction outside the confines of the social-psychological laboratory.

Conformity is one of the most obvious and basic characteristics of social behavior. In extensive laboratory research it has been found that subjects will conform to an objectively incorrect group opinion, and that the degree of conformity is dependent on such factors as the size of the consensus group, the perceived expertness of the group, the presence of other nonconforming members in the group, and the difficulty of the objective judgment being made by the group. The mere order in which the members of the group express their judgments may have dramatic effects on the subject's rating of their personality and intellect.

Group processes comprise an important aspect of the study of social behavior. Groups have been classified in many ways. For example, nominal groups are collections of individuals working relatively independently, while true groups tend to work together for some common objective. True groups can be further differentiated as task oriented or process oriented, or as cooperative or competitive. In general cooperative groups tend to be more productive than competitive groups. Groups can differ in terms of distribution of authority, as in democratic leadership versus authoritarian leadership. Finally, leaders themselves may be task oriented or process oriented; their effectiveness for a particular type of group depends to some extent on the group goal. The qualities of leadership may well depend more on the circumstances than on any innate qualities of the leader.

Perhaps one of the most thoroughly investigated areas of social psychology is the process of attitude acquisition and change. Important influences on attitude change are the credibility of the source of a communication, the plausibility of the arguments employed, the degree of coorientation of audience and speaker. Other factors are the strength, complexity, and saliency of preexisting attitudes.

BIBLIOGRAPHY

Adams, R. N. 1960. An inquiry into the nature of the family. In G. E. Dole and R. L. Carneiro (Eds.), *Essays in the science of culture in honor of Leslie A. White.* New York: Crowell.

Ardrey, R. 1966. *The territorial imperative.* New York: Atheneum. (Copyright 1966 by Robert Ardrey; reprinted by permission of the publisher.)

Aronson, E., and Lindner, D. 1965. Gain and loss of esteem as determinants of interpersonal attraction. *J. Exp. Soc. Psych.,* 1, 156–171.

Asch, S. E. 1951. Effects of group pressure on the modification and distortion of judgments. In H. Geutzkow (Ed.), *Groups, leadership, and men.* Pittsburg: Carnegie. (By permission.)

Bandura, A., and Walters, R. H. 1963. Aggression. In H. W. Stevenson (Eds.), *Child psychology.* Chicago: University of Chicago Press.

Bavelas, A., Hastorf, A. H., Gross, A. E., and Kite, W. A. 1965. Experiments on the alteration of group structure. *Exp. Soc. Psychol.,* 1, 55–70.

Beach, F. A. 1947. Evolutionary changes in the physiological control of mating behavior of mammals. *Psychol. Rev.,* 54, 297–315.

Beach, F. A. 1969. Locks and beagles. *Amer. Psychol.,* 24, 921–949, 971–989.

Bell, R. R. 1966. *Premarital sex in a changing society.* Englewood Cliffs, N.J.: Prentice-Hall.

Berkowitz, L. 1968. Impulse aggression and the gun. *Psychol. Today,* 2, 18–23.

Bernard, J. 1968. *The sex game.* Englewood Cliffs, N.J.: Prentice-Hall.

Bowlby, J. 1969. *Attachment and loss,* Vol. I, *Attachment.* New York: Basic Books.

Bridges, K.M.B. 1932. Emotional development in early infancy. *Child. Devel.,* 3, 324–341.

Bugelski, B. R. 1956. *The psychology of learning.* New York: Henry Holt. (By permission.)

Byrne, D. 1961. Interpersonal attraction and attitude similarity. *J. Abnorm. Soc. Psychol.,* 62, 713–715.

Cairns, R. B. 1966. Attachment behavior of mammals. *Psychol. Rev.,* 73, 409–426.

Caldwell, B. M. 1964. The effects of infant care. In L. W. Hoffman and M. L. Hoffman (Eds.), *Review of child development research.* Vol. I. New York: Russell Sage.

Chamove, A., Harlow, H. F., and Mitchell, G. 1967. Sex differences in the infant-directed behavior of preadolescent rhesus monkeys. *Child Devel.,* 38, 329–335.

Cooper, J. B., and Blair, M. A. 1959. Parent evaluation as a determiner of ideology. *J. Genetic Psychol.,* 94, 93–100.

Cooper, J. B., and McGaugh, J. L. 1963. *Integrating principles of social psychology.* Cambridge: Schenkman. (By permission.)

Darwin, C. 1872. *The expression of the emotions in man and animals.* New York: D. Appleton. Reprinted in 1965 by University of Chicago Press.

Deets, A. C. 1969. The effects of twinship on the interactions between rhesus monkey mothers and infants. (Master's thesis, University of Wisconsin) Madison, Wisc.

Deutsch, M. 1949. An experimental study of the effects of cooperation and competition upon group processes. *Human Relations,* 2, 199–231.

DeVore, I. 1963. Mother-infant relations in free-ranging baboons. In H. L. Rheingold (Ed.), *Maternal behavior in mammals.* New York: Wiley.

Dewey, J. 1922. *Human nature and conduct.* New York: Henry Holt.

Elias, J., and Gebhard, P. 1969. Sexuality and sexual learning in children. *Phi Delta Kappan,* 50, 401–405.

Erikson, E. H. 1950. *Childhood and society.* New York: Norton.

Festinger, L. 1954. A theory of social comparison processes. *Human Relations,* 7, 117–140.

Festinger, L., Schachter, S., and Back, K. 1950. *Social pressures in informal groups: A study of human factors in housing.* New York: Harper. (By permission of the publisher.)

Freud, S. 1949. *An outline of psychoanalysis.* New York: Norton.

Gari, J. E., and Scheinfeld, A. 1968. Sex differences in mental and behavioral traits. *Genetic Psychol. Monog.,* 77, 169–299.

Goodenough, F. L. 1931. *Anger in young children.* Minneapolis: University of Minnesota Press.

Groos, K. 1901. *The play of man.* New York: D. Appleton-Century.

Gunther, M. 1961. Infant behaviour at the breast. In B. M. Foss (Ed.), *Determinants of infant behavior.* Vol. I. London: Methuen.

Hall, G. S. 1920. *Youth.* New York: D. Appleton-Century.

Hamburg, D. A., and Lunde, D. T. 1966. Sex hormones in the development of sex differences in human behavior. In E. E. Maccoby (Ed.), *The development of sex differences.* Stanford: Stanford University Press.

Hansen, E. W. 1966. The development of maternal and infant behavior in the rhesus monkey. *Behaviour,* 27, 107–149.

Harlow, H. F. 1958. The nature of love. *Amer. Psychol.,* 13, 673–685.

Harlow, H. F. 1969. Age-mate or peer affectional system. In D. S. Lehrman, R. A. Hinde, and E. Shaw (Eds.), *Advances in the study of behavior.* Vol. 2. New York: Academic Press.

Harlow, H. F., and Griffin, G. 1965. Induced mental and social deficits in rhesus monkeys. In S. F. Asler and R. E. Cooke (Eds.), *The biosocial basis of mental retardation.* Baltimore: Johns Hopkins Press.

Harlow, H. F., and Harlow, M. K. 1965. The affectional systems. In A. M. Schrier, H. F. Harlow, and F. Stollnitz (Eds.), *Behavior of nonhuman primates.* Vol. II. New York: Academic Press.

Harlow, H. F., Harlow, M. K., and Hansen, E. W. 1963. The maternal affectional system of rhesus monkeys. In H. L. Rheingold (Ed.), *Maternal behavior in mammals.* New York: Wiley.

Harlow, H. F., Joslyn, W. D., Senko, M. G. and Dopp, A. 1966. Behavioral aspects of reproduction in primates. *J. Animal Sci.,* 25, 49–67.

Harlow, H. F., and Zimmermann, R. R. 1959. Affectional responses in the infant monkey. *Science,* 130, 421–432.

Harlow, M. K., and Harlow, H. F. 1966. Affection in primates. *Discovery,* 27, 11–17.

Hastorf, A. H. 1966. Personal communication.

Hess, E. H. 1959. Imprinting. *Science,* 130, 133–141.

Hirschberg, G., and Gilliland, A. R. 1942. Parent-child relationships and attitudes. *J. Abnorm. Soc. Psychol.,* 37, 125–130.

Horowitz, E. L. 1936. The development of attitude toward the negro. *Arch. Psychol.,* 28, 194. (By permission.)

Hovland, C. I., and Weiss, W. 1951. The influence of source credibility on communication effectiveness. *Publ. Opin. Quart.,* 15, 635–650. (By permission.)

Hovland, C. I., Lumsdaine, A. A., and Sheffield, F. D. 1949. Experiments on mass communication. In *Studies in social psychology in world war II.* Vol. 3. Princeton: Princeton University Press.

Janis, I. L., and Feshback, S. 1953. Effects of

fear-arousing communications. *J. Abnorm. Soc. Psychol.*, 48, 78–92. (Copyright 1953 by the American Psychological Association, reproduced by permission.)

Jensen, G. D. 1965. Mother-infant relationship in the monkey *Macaca nemestrina:* Development of specificity of maternal response to own infant. *J. Comp. Physiol. Psychol.*, 59, 305–308.

Jersild, A. T., and Holmes, F. B. 1935. Methods of overcoming children's fears. *J. Psychol.*, 1, 75–104.

Jersild, A. T., Mackey, F. V., and Jersild, C. L. 1933. Children's fears, dreams, wishes, daydreams, likes, dislikes, pleasant and unpleasant memories. *Child. Devel. Monogr.*, No. 12.

Jones, E. E., and Gerard, H. B. 1967. *Foundations of social psychology.* New York: Wiley.

Jones, H. E. 1954. The environment and mental development. In R. Carmichael (Ed.), *Manual of child psychology.* New York: Wiley.

Jones, N.G.B. 1967. An ethological study of some aspects of social behaviour of children in nursery school. In D. Morris (Ed.), *Primate ethology.* Chicago: Aldine.

Kagan, J., and Moss, H. A. 1962. *Birth to maturity: A study in psychological development.* New York: Wiley.

Kaufman, I. C., and Rosenblum, L. A. 1969. The waning of the mother-infant bond in two species of macaque. In B. M. Foss (Ed.), *Determinants of infant behavior.* Vol. IV, London: Methuen.

Kessen, W., and Mandler, G. 1961. Anxiety, pain and the inhibition of distress. *Psychol. Rev.*, 68, 396–404.

Krech, D., and Crutchfield, R. S. 1948. *Theory and problems of social psychology.* New York: McGraw-Hill. (Used with permission of McGraw-Hill Book Company.)

Latane, B. 1966. Studies in social comparison. *J. Exp. Soc. Psychol.*, Suppl. 1.

Lewin, K., Lippitt, R., and White, R. K. 1939. Patterns of aggressive behavior in experimentally created social climates. *J. Soc. Psychol.*, 10, 271–301.

Lippitt, R. 1940. An experimental study of the effect of democratic and authoritarian group atmospheres. *Univ. Iowa Stud.*, 16(3), 43–198.

Lorenz, K. 1952. *King Solomon's ring.* London: Methuen.

Louttet, C. M. 1927. Reproductive behavior of the guinea pig. *J. Comp. Psychol.*, 7, 247–263.

Lumsdaine, A. A., and Janis, I. L. 1953. Resistance to "counterpropaganda" produced by one-sided and two-sided "propaganda" presentations. *Publ. Opin. Quart.*, 17, 311–318.

Maccoby, E. 1966. *The development of sex differences.* Stanford: Stanford University Press.

Mandler, G. 1964. The interruption of behavior. In D. Levine (Ed.), *Nebraska symposium on motivation.* Lincoln: University of Nebraska Press.

Masters, W. M., and Johnson, V. E. 1970. *Human sexual inadequacy.* Boston: Little, Brown.

McCammon, R. W. 1965. Are boys and girls maturing at earlier ages? *Amer. J. Publ. Health*, 55, 103–106.

McGraw, M. B. 1935. *Growth: A study of Johnny and Jimmy.* New York: D. Appleton-Century.

Milgram, S. 1965. Some conditions of obedience and disobedience to authority. *Human Relations*, 18, 57–76.

Mitchell, G. D. 1969. Paternalistic behavior in primates. *Psychol. Bull.*, 71, 399–417.

Money, J. 1970. Determinants of human sexual behavior. In A. M. Freedman, H. I. Kaplan, and H. S. Kaplan (Eds.), *Comprehensive textbook of psychiatry* (2nd ed.). Baltimore: Williams and Wilkins.

Newcomb, T. M. 1961. *The acquaintance process.* New York: Holt, Rinehart and Winston.

Piaget, J. 1967. *Six psychological studies.* New York: Random House.

Pratt, C. L., and Sackett, G. P. 1967. Selection of social partners as a function of peer contact during rearing. *Science*, 155, 1133–1135. (Copyright 1967 by the American Association for the Advancement of Science, by permission.)

Radke, M., and Sutherland, J. 1949. Children's concepts and attitudes about minority and majority American groups. *J. Educ. Psychol.*, 40, 449–468.

Ramsey, G. V. 1943. The sexual development of boys. *Amer. J. Psychol.*, 56, 217–233.

Rank, O. 1952. *The trauma of birth.* New York: Basic Books.

Rheingold, H. L., and Eckerman, C. O. 1970. The infant separates himself from his mother. *Science,* 1968, 78–83.

Robson, K. S., and Moss, H. A. Bethesda, Maryland: Child Research Branch, National Institute of Mental Health. Unpublished findings.

Roe, A. 1952. *The making of a scientist.* New York: Dodd, Mead.

Rosenblith, J. F. 1949. A replication of "some roots of prejudice." *J. Abnorm. Soc. Psychol.,* 44, 470–489.

Roth, P. 1969. *Portnoy's complaint.* New York: Random House.

Sackett, G. P. 1966. Monkeys reared in visual isolation with pictures as visual input: Evidence for an innate releasing mechanism. *Science,* 154, 1468–1472.

Salk, L. 1960. Effects of the normal heartbeat sound on the behavior of the newborn infant. Implications for mental health. *World Mental Health,* 12, 168–175.

Saranson, S. B. 1960. *Anxiety in elementary school children: A report of research by Seymour B. Saranson and others.* New York: Wiley.

Schachter, S. 1959. *The psychology of affiliation.* Stanford: Stanford University Press. (By permission of the publisher. Copyright © 1959 by the Board of Trustees of the Leland Stanford Junior University.)

Schaffer, H. R., and Emerson, P. E. 1964*a*. The development of social attachments in infancy. *Monogr. Soc. Res. Child Devel.,* 29(3), 1–77.

Schaffer, H. R., and Emerson, P. E. 1964*b*. Patterns of response to physical contact in early human development. *J. Child Psychol. Psych.,* 5, 1–13.

Sears, R. R., Whiting, J.W.M., Nowlis, V., and Sears, P. S. 1953. Some child rearing antecedents of aggression and dependency in young children. *Genetic Psychol. Monogr.,* 47, 135–234.

Spencer, H. 1873. *The principles of psychology.* New York: D. Appleton-Century.

Spitz, R. A. 1946*a*. Anaclitic depression. In *The psychoanalytic study of the child.* Vol. II. New York: International Universities Press.

Spitz, R. A. 1950. Anxiety in infancy: A study of its manifestations in the first year of life. *Intern. J. Psycho-Anal.,* 31, 138–143.

Suomi, S. J., Harlow, H. F., and Lewis, J. K. 1970. Effect of bilateral frontal lobectomy on social preferences of rhesus monkeys. *J. Comp. Physiol. Psychol.,* 70, 448–453. (Copyright 1970 by the American Psychological Association, reproduced by permission.)

Taylor, C. W. (Ed.). 1964. *Creativity: Progress and potential.* New York: McGraw-Hill.

Thompson, T. I. 1963. Visual reinforcement in Siamese fighting fish. *Science,* 141, 55–57.

Tuddenham, R. D., and MacBride, P. D. 1959. The yielding experiment from the subject's point of view. *J. Personal.,* 27, 259–271. (By permission.)

Valins, S. 1966. Cognitive effects of false heart-rate feedback. *J. Personal. Soc. Psychol.,* 4, 400–408.

Valins, S., and Ray, A. A. 1967. Effects of cognitive desensitization on avoidance behavior. *J. Personal. Soc. Psychol.,* 7, 345–350.

Van Lawick-Goodall, J. 1967. Motor-offspring relationships in free-ranging chimpanzees. In D. Morris (Ed.), *Primate ethology.* Chicago: Aldine.

Van Wagenen, G. 1950. The monkey. In E. J. Farris (Ed.), *The care and breeding of laboratory animals.* New York: Wiley.

Walster, E., Aronson, V., Abrams, D., and Rottman, L. 1966. Importance of physical attractiveness in dating behavior. *J. Personal. Soc. Psychol.,* 4, 508–516.

Watson, J. B., and Rayner, R. 1920. Conditioned emotional reactions. *J. Exp. Psychol.,* 3, 1–14.

INDEX